Principles of Food Science and Nutrition

Principles of Food Science and Nutrition

Dr. Vijayalakshmi, D.
Professor (HAG) and Head
Department of Food Science and Nutrition
UAS, GKVK, Bengaluru-560 065

Dr. Usha Ravindra
Associate Professor (FSN)
Department of Food Science and Nutrition
UAS, GKVK, Bengaluru-560 065

Dr. Shamshad Begum, S.
Assistant Professor (FSN)
Department of Food Science and Nutrition
UAS, GKVK, Bengaluru-560 065

International Books & Periodical Supply Service
(Publisher of Scientific Books)
38, NISHANT KUNJ, PITAM PURA MAIN ROAD
DELHI-110034 (India)
Phone : 011-27352078, Mobile : 09810146811, E-mail : hkjain1975@yahoo.com

Published by :

International Books & Periodical Supply Service

(Publisher of Scientific Books)
38, NISHANT KUNJ, PITAM PURA MAIN ROAD
DELHI-110034 (India)
Phone : 011-27352078, Mobile : 09810146811, E-mail : hkjain1975@yahoo.com

ISBN 978-81-935695-9-7

Composed, Designed & Printed in India

University of Agricultural Sciences, Bengaluru
Gandhi Krishi Vigyan Kendra, Bangalore-560 065

Website : www.uasbangalore.edu.in
E-mail : vc@uasbangalore.edu.in
vcuasb1964@gmail.com
srprasad1989@yahoo.co.in

Off. : +91 080 2333 2442
Mob. : +91 94498 66900
Fax : +91 080 2333 0277

Dr. S. Rajendra Prasad
Vice Chancellor

Foreword

Food Science and Nutrition has a very important role in providing safe, nutritious and wholesome food by reducing food losses to maximize efficiency of resource use with a balanced approach. 'Principles of Food Science and Nutrition' in a broader sense imply production, processing, preparation, preservation, and metabolizing food. It explains about how nutrients and other food components illustrate basic food chemistry, measurements and functionalities, role of macro and micronutrients and also the effect of deficiency on humans. It has various methods of food processing, preservation and packaging using traditional and modern technologies with emphasis on food safety and quality control. To be healthy, we must all make sure that we are consuming balanced food with right amount of nutrients. Our diet should be rich in wholesome foods, such as cereals, millets, pulses, nuts and oilseeds, fruits and vegetables, fleshy foods to get adequate nutrients to sustain sound health with vigour.

India has made rapid strides in improving rates of under and malnutrition. Between 2006 and 2016, stunting in children below five years declined from 48 per cent to 38 per cent. Yet, India continues to have one of the world's highest child undernutrition rates, impacting the child's health and development, performance in school and productivity in adult life. Also, Global hunger is on the rise again, affecting 815 million people in 2016, as per the UN annual report on world food security and nutrition. In India, the proportion of the population that is undernourished is 22 per cent, the prevalence of underweight in children below five years is 43.5 per cent and proportion of children dying before the age of five is 6.9 per cent. The global Hunger

Index places India 67th in ranking with a score of 24.1 far below China, Sri Lanka, Pakistan and Nepal. According to FAO, about 25 per cent of world's undernourished live in India. The government has large food security and anti-poverty programmes but there are critical gaps in terms of inclusion and exclusion errors. Women, children and girls are particularly vulnerable and disadvantaged. The National Food Security Act (NFSA), 2013 aims to ensure food and nutrition security for the most vulnerable through its associated schemes and programmes making access to food a legal right.

With this background, a focused book on Principles of Food Science and Nutrition is compiled to get the precise text as per 'V Deans' committee's recommendations' of ICAR, New Delhi, Government of India. It presents information covering the concepts of food science and nutrition, composition, classification, functions and sources, food microbiology. food processing and preservation, nutritional disorders, government prophylaxis programmes, energy metabolism, balanced diet and menu planning along with new trends. This is a maiden attempt from the faculty to bring out the publication on Principles of Food Science and Nutrition for the benefit of students and others interested. I congratulate the authors, Dr. Vijayalakshmi, Dr. Usha Ravindra and Dr. Shamshad Begum, S. for their sincere and dedicated efforts in bringing out this endeavour as a instructional material meticulously and I wish them all the success.

(S. Rajendra Prasad)

Preface

Food is any substance consumed to provide nutritional support for an organism. It is usually of plant or animal origin, and contains essential nutrients, such as carbohydrates, fats, proteins, vitamins and minerals. Despite green revolution which paved the way for better food security by quadrupling the production, still there is a huge disconnect between Agriculture and Nutrition. It is the time we pay attention for the enhancement of food and nutritional security of people in the community. As we know malnutrition affects more people in India, particularly children below the age of three, than anywhere else in the world. **An attempt has been made to meet the needs of the students of Indian agricultural universities pursuing courses in graduation in Agriculture/ Horticulture/Biotechnology/Sericulture/Food Science and Technology.** A focused book on Principles of Food Science and Nutrition is compiled to get the precise text as per 'V Deans' committee's recommendations' of ICAR, New Delhi, Government of India. The desire to write 'Principles of Food Science and Nutrition' had been inculcated while teaching to students. The literature on Principles of Food Science and Technology is extensive. However, an attempt has been made to provide as much knowledge as possible in a nutshell in the following 9 chapters covering all the contents of the syllabus information comprised of the introduction to food science and concepts related to nutrition, food composition, classification, functions and nutrients like carbohydrates, protein, fats, energy, water, vitamins, minerals, bioactive compounds. It also includes food microbiology, principles, methods of food processing and preservation, nutritional disorders in detail along with prophylaxis programmes of government for nutritional anemia, vitamin A and iodine deficiency including ICDS, PDS and mid-day meals. Also energy metabolism, balanced diet, modified diet and menu planning and new trends in Food Science & Nutrition, such as irradiation, extrusion, aseptic processing, transfatty acids, nutraceuticals, pre and probiotics, phytochemicals and nutrition labelling. The book gives a comprehensive knowlegde about foods. We wish to convey our sincere and grateful thanks to our honourable Dean(Agri.), Dr. S. Rajendra Prasad, College of Agriculture, University of Agricultural Sciences, GKVK, Bengaluru-560 065 for his motivation, constant encouragement and support for this endeavour.

Dr. Vijayalakshmi, D.
Dr. Usha Ravindra
Dr. Shamshad Begum, S.

List of Tables

List of Figures

Contents

Chapter 1

Concepts of Food Science

Food Science is a systematic study of the nature of food materials and the scientific principles underlying their modification, preservation, and spoilage. To understand food science, the basic concepts of physics, chemistry, mathematics and biology and their applications, *i.e.,* biochemistry, microbiology, and food technology, are necessary in order to prepare, package, store, and serve wholesome, high quality products. All foods are chemical compounds which undergo various chemical reactions at all stages from production to consumption. These reactions are based on the laws of chemistry. Many processes used while preparing food involve physical changes apart from chemical changes. Processes used while preparing food results in physical changes apart from chemical changes. Matter exists in three states: solid, liquid, and gas.

Terminologies Commonly used in Human Nutrition

Food Science: A study of the physical and chemical constituents of food and the scientific principles underlying their modification, preservation, and spoilage.

Staple food: A food is known as a staple food when it is eaten regularly, especially by the poorer population sectors of a country or community, and in large enough quantities to furnish a substantial proportion of the total energy supply.

Food intake : All the food and drink an individual consumes in a given length of time, usually a 24-hrs period is called that person's food intake.

Diet: A diet consists of the total solid and liquid foods consumed by an individual or a population group. It is a sum of all meals consumed during a 24 hours period.

Dietary habit: The way we generally choose our daily food becomes our dietary habit.

Balanced diet: When the diet meets all the nutritional requirements of the body, it is called a balanced diet.

Recommended Dietary Allowance (RDA): The Recommended Dietary Allowance refers to the amount of energy (Calories) and nutrients which will maintain a good state of nutrition in healthy people of all ages. The allowance includes a margin of safety to cover individual differences.

Food preservation: Most foods can be safely saved for future out-of season use through preservation by one or more methods. Some foods may be kept simply by storing in a clean dry place; others may require salting, drying, smoking, curing, fermenting or heating and canning, bottling or pickling, etc. Use of sanitary food preservation methods will increase foods available for consumption, both at home and for sale.

Food technology: Application of the principles of food science to the preservation, processing, packaging, storage, and transportation of food materials.

Food Processing: Conversion of a produce into a product with added value by different basic processing of dehydration, pulp extraction, powder making, cooking, frying, etc. to get a condition that is more finished, stored or marketable, e.g. Cereal flour, Turdal, Fruit Jam.

Food Hygiene: All conditions and measures necessary to ensure the safety and suitability of food at all stages of food chain.

Food Safety: Assurance that food will not cause harm to the consumer when it is prepared and /or eaten according to its intended use.

Food Suitability: Assurance that food is acceptable for human consumption according to its intended use.

Quality Assurance: It refers to the measures taken to make the product safe or assuring the quality at all stages.

Quality Control: May be defined as the maintenance of quality at all level and tolerance acceptable to the buyer, while minimizing the cost for vendor. It is nothing but a testing of the product for quality.

a) Measurements

Length

- The unit for measuring length is the metre (m).
- Length is measured using a measuring tape or ruler.
- One thousand metres (1,000 m) = one kilometre (km)
- A metre is divided into hundred parts. Each part is called a centimetre (cm)
- A centimetre is made up of ten smaller parts called millimetre (mm).

Volume

- Volume and capacity is measured in litres. A litre is made up of 10 decilitres (dl). Each decilitre is made up of 10 centilitres (cl), A centilitre is made up of 10 millilitres (1,000 ml).
- Most measuring cups and jugs are marked in millilitres and litres. The capacity of cups and spoons are listed below:

  1 table spoon = 15 ml

  1 tea spoon = 5 ml

  1 breakfast cup = 240 ml

  1 coffee cup = 100-120 ml

  A tea cup = 150-180 ml

  1 water glass = 280-300 ml

The volume of solids that is not greatly affected by water can be measured by the water displacement method. Solids are immersed in the displacement can and the volume of water displaced, equal to the volume of the solid, is noted.

The seed method is used to measure the volume of cake and bread. A large tin box is filled to the brim with seeds and the volume of seeds required to fill the box is measured in a measuring cylinder. The cake of which the volume is to be measured is placed in the empty tin and covered with seeds. The volume of seeds remaining after covering the cake is equal to the volume of the cake.

Weight or Mass

- Weight is the pull experienced on the body by the earth's force of gravity. Mass is the amount of matter contained in a known volume of substance. Mass always remains constant but weight may change in different parts of the world because the force of gravity varies from place to place.
- Weight is measured on a weighing scale. The kilogram is the unit for measuring weight and is made up of one thousand smaller parts called grams.

  1 kilogram (kg) = 1,000 grams (g)

  Each gram is further divided into one thousand smaller parts called milligrams (mg).

  1 g = 1,000 mg

  Each milligram is further divided into 1,000 micrograms (μg).

  1mg = 1,000 Microgram (μg)

  From the above we conclude that

  1 kg = 1,000,000 mg and a measure of 1 ppm means 1 mg in 1 kg of a substance.

Temperature

- Temperature is a physical quantity expressing hot and cold. Temperature is measured with a thermometer, historically calibrated in various temperature scales and units of measurement. The most commonly used scales are the Celsius scale, denoted in °C (informally, degrees centigrade), the Fahrenheit scale (°F), and the Kelvin scale. The Kelvin (K) is the unit of temperature in the International System of Units (SI), in which temperature is one of the seven fundamental base quantities.
- Temperature refers to the relative hotness or coldness of a substance compared with melting ice at 0 ^{0}C and boiling water at 100 ^{0}C. Thermometers are used to measure temperature.
- Temperature is measured either in the Celsius or Centigrade scale (^{0} C) or in the Fahrenheit scale (^{0} F). Each scale has two fixed points:

1. Melting point of ice (0^0C or 32 ^{0}F)
2. Boiling point of water (100^0C or 212^0F).

Heat is a form of energy needed to carry out work. Energy is the capacity of doing work. Energy is presented in two forms:

1. Potential energy or stored energy, such as the energy stored in a bar of chocolate.
2. Kinetic energy or active energy in motion, such as when a person is walking.

Energy is present in many forms. Heat is one form of energy. Solar energy, electrical energy and chemical energy are some of the others. Heat energy is measure in units called joules and the energy present in food is measured in kilocalories. One kilocalorie is made up of 1,000 calories.

1 kilocalorie (kcal) = 4.2 kilojoules (kj)

1 calorie = 4.2 joules

Temperature is important in all fields of natural science, including physics, chemistry, Earth science, medicine, and biology, as well as most aspects of daily life.

b) Density

Density is the relationship between the weight and volume of a substance expressed as,

$$\text{Density} = \frac{\text{Weight in kg}}{\text{Volume in m}^2}$$

It is expressed in kilograms per cubic metre and is used to compare the heaviness or lightness of different foods.

A fruit cake has a greater density as compared to a sponge cake. The density of liquids is measured in g/cm^3. Water has a density of 1g/cm^3.

Relative density

Relative density (RD) is the ratio of the mass of a known volume of a substance to the mass of the same volume of water. It tells us the number of times the volume of a substance is heavier or lighter than an equal volume of water. If the RD of a volume of lead is 11, it means that is eleven times as heavy as an equal of water.

$$\text{Relative density} = \frac{\text{Mass or weight of a substance}}{\text{Weight of equal volume of water}}$$

A hydrometre is used to measure the relative density of different liquids. Saccharometer for sugar, Salinometer for brine solution, Lactometer for purity of milk, Refractometer for TSS in jams and jellies.

c) Colloidal system and phase change-liquid, gaseous and solid phase

A colloid is one of the three primary types of mixtures, with the other two being a solution and suspension. A colloid is a solution that has particles ranging between 1 and 1000 nanometres in diameter, yet are still able to remain evenly distributed throughout the solution. These are also known as colloidal dispersions because the substances remain dispersed and do not settle to the bottom of the container. In colloids, one substance is evenly dispersed in another. The substance being dispersed is referred to as being in the dispersed phase, while the substance in which it is dispersed is in the continuous phase.

- An intimate mixture of two substances, one of which, called the dispersed phase (or colloid), is uniformly distributed in a finely divided state through the second substance, called the dispersion medium (or dispersing medium).
- The dispersion medium may be a gas, a liquid, or a solid and the dispersed phase may also be any of these, with the exception of one gas in another. A system of liquid or solid particles colloidally dispersed in a gas is called an aerosol. A system of solid substance or water-insoluble liquid colloidally dispersed in liquid water is called a hydrosol. There is no sharp line of demarcation between true solutions and colloidal systems or between mere suspensions and colloidal systems. When the particles of the dispersed phase are smaller than about 10^{-3} ìm in diameter, the system begins to assume the properties of a true solution; when the particles dispersed are much greater than 1 ìm, separation of the dispersed phase from the dispersing medium becomes so rapid that the system is best regarded as a suspension.

A common method of classifying colloids is based on the phase of the dispersed substance and what phase it is dispersed in. The types of colloids include sol, emulsion, foam, and aerosol.

1. **Sol** is a colloidal suspension with solid particles in a liquid.
2. **Emulsion** is between the two liquids.
3. **Foam** is formed when many gas particles are trapped in a liquid or solid.
4. **Aerosol** contains small particles of liquid or solid dispersed in a gas.

When the dispersion medium is water, the colloidal system is referred to as a hydrocolloid. The particles in the dispersed phase can take place in different phases depending on how much water is available. For example, Jelly powder mixed with water creates a hydrocolloid.

d) pH (Potential Hydrogen)

When an acid is diluted with water, it dissociates into hydrogen ions and acid radical ions. The term pH (Hydrogen ion concentration) is used to express the degree of acidity or alkalinity of a food. It is defined as the negative logarithm to the base 10 of the hydrogen ion concentration. Higher the hydrogen ion concentration lower will be the pH and vice versa.

Table 1: Types of colloidal system

Systems	Disperse Phase	Continuous Phase	Products
Sol	Solid	Liquid	Uncooked custard, unset jelly
Gel	Liquid	Liquid	Jelly, Jam
Emulsion	Liquid	Liquid	Mayonnaise, milk
Solid emulsion	Liquid	Liquid	Butter, margarine
Foam	Gas	Liquid	Whipped cream, whisked egg white
Solid Foam	Gas	Liquid	Bread, cake, ice cream

Buffers

They are defined as solutions that can resist a change of pH on addition of acids or alkalis but within limits. These solutions are made up of a weak acid and one of its salts or a weak base and one of its salts.

Application of pH

Preparation of Jam - addition of citric acid to adjust the pH to slightly acidic.

Retaining bright colour in green leafy vegetables - addition of Sodium bicarbonate.

e) Osmosis

Osmosis is the passage of water from a weak solution through a semi-permeable membrane.

When raisins are soaked in a cup of water for some time, the raisins swell because water from the cup enters the raisins. Similarly, if raisins are placed in a concentrated sugar solution, they shrivel up after some time because water from the raisins passes into the sugar solution because of osmosis.

Plant and animal cell membranes act as semi-permeable membranes and selectively permit water and electrolytes to enter or leave the cell.

Applications of osmosis

1. Osmosis plays an important role in food processing and preservation to retain the original shape and size of canned fruits in syrup and of vegetables in pickles.
2. The freshness of fruits and vegetables depends on the osmotic pressure in the cells. Salads lose their crisp crunchy texture and become limp if salt and sugar is sprinkled much in advance. Lettuce leaves can be revived by immersing them in chilled water.

f) Surface tension

Surface Tension is a force experienced on the surface of a liquid. It is caused by cohesion, i.e., a force that causes the molecules of a substance to be attracted to one another.

The molecules of a liquid that are below the surface are pulled by cohesive forces from all directions. But the molecules at the surface behave differently because they are only pulled downwards or sideways. This downward or sideways attraction causes a constant pull on the surface molecules which makes the liquid behave as if it is covered by a thin elastic film, For example, the surface of water can support needles if they are placed carefully.

Because of surface tension, drops of liquid take a spherical shape, which has the smallest possible surface are, e.g., dew drops.

Surface tension causes liquids to rise in a thin tube (capillary tube when the tube is dipped in liquid. This property of liquids is important in many food systems and in the action of detergents.

Surface tension is also defined as the force of attraction which exists between liquid and solid surfaces.

Applications of surface tension

1. Release agents help prevent the paper lining the tin from sticking to the cake. They contain silicone compounds.
2. Silicones have a property of lowering the surface tension and is added to wood polishes to allow the polish to spread easily.
3. Non-stick cookware is coated with polytetrafluoroethane plastic or silicone to prevent attraction between the food and pan.

g) Evaporation

Evaporation is a change of state from liquid to gas which takes place continuously from the surface of a liquid. Volatile liquids vaporize easily, e.g., petrol and acetone. Non-volatile liquids, such as oils, evaporate very gradually Evaporation is faster when there is breeze and low humidity in the air as well as a large surface area and high temperature.

Applications of evaporation

1. Bread and cake if left uncovered, hardens and becomes stale because of loss of moisture. This can be prevented by storing food in covered tins.
2. Cooking in shallow uncovered pans will cause greater evaporation and are used for preparing mawa from milk.

3. Milk powder is prepared by dehydration or spray drying in which water from milk is removed by circulating hot air.

h) Boiling point

Boiling is the use of heat to change a substance from a liquid to a gas. The change takes place throughout the body of the liquid at a definite temperature.

Like the melting point, the boiling point of a pure substance is always constant. It changes if impurities or dissolved substances are present or by changes in atmospheric pressure/Pure water boils at 100^{0}C.

Applications of boiling point

1. Boiling vegetables in salted water increase the boiling point above 100°C.
2. In sugar cookery, the boiling points of sugar solutions is noted at various stages so that fondant, fudge, toffee, and caramel can be prepared.

i) Melting point

Melting or fusion is the change of state from a solid to a liquid.

The temperature at which a solid melts and turns into a liquid is called its melting point. The melting point of fats depends on the percentage of saturated long chain fatty acids present in it.

The melting point for any chemical is fixed and is used to measure the purity of a substance. It is lowered by adding other substances.

Applications of melting point

1. Ice has a melting point of 0^{0}C. If adequate sodium chloride is added to ice, the melting point falls to -18^{0}C. This lowering of melting point is used in the setting of ice cream.
2. Fat is removed from adipose tissue of animals by a process called rendering, which is based on the melting point. Boiling water or dry heat is used to liberate the oil from the fat cells.

j) Humidity

Humidity refers to the presence of water vapour in the air. Water vapour is produced by respiration of plants and animals, evaporation from food during cooking and from water bodies, from rain during the monsoons, etc.

In catering establishments, moisture in the air is quite high because of large volumes of steam from boilers, from cooking food, from dishwashers and laundry processes, and respiration and perspiration of people in a confined area.

A humid atmosphere causes discomfort, headache, and tiredness.

The humidity of the air is measured with the help of a hygrometer. This instrument depicts the percentage of water vapour in the air. It is a ratio between the amount of water vapour which air could hold and what it actually holds at the same temperature. Humidity of 60-70 per cent is considered normal and does not cause discomfort or undue spoilage of food.

Applications of humidity

1. Spoilage organisms multiply and spores germinate at high moisture levels in the atmosphere.
2. Processed foods are prevented from drying up by adding substances with hygroscopic properties called humectants. Glycerine and sorbitol are used as humectants in jam.

k) Water activity (aw)

Water activity represents the ratio of water vapour pressure of the food to the water vapour pressure of pure water under the same conditions and it is expressed as a fraction.

$aw = P/p_o$

P = Vapour pressure of the water in the solution

P_o = Vapour pressure of the pure water at the same temperature.

The Water activity scale extends from zero(bone dry) to 1.0 (pure water) but most foods have a water activity level in the range of 0.2 (very dry fruits) to 0.99 (moist fresh foods). The Water activity is equilibrium relative humidity (ERH). The Water activity of a food is not the same thing as its moisture content.

Table 2: Water activity of foods

Food stuff	Water activity
Fresh meat and fish	0.99
Bread	0.95
Aged cheddar	0.85
Jams and Jellies	0.80
Plum pudding	0.80
Dry fruit	0.60
Biscuits	0.30
Milk powder	0.20
Instant coffee	0.20

❑❑❑

Chapter 2

Food Composition, Classification, Functions and Food Sources

1. Carbohydrates

Carbohydrates are hydrates of carbon which are one of the four major classes of biomolecules and play several important roles in all life forms. Carbohydrates are synthesized by all green plants using solar energy, water from the soil and carbon dioxide from the air. This complex process is called photosynthesis. Carbohydrates contain carbon, hydrogen and oxygen. Carbohydrates are the primary fuel for our body, cereals are our staple food. They are a rich source of carbohydrates. In the Indian diet, about 65 to 80 per cent energy is provided by carbohydrates.

a) Classification

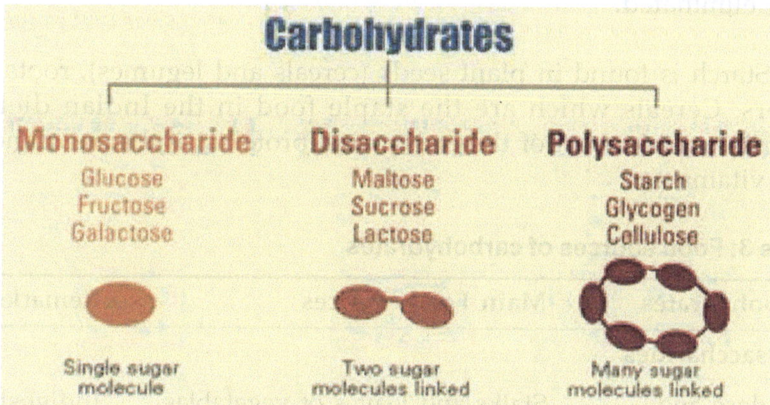

Fig. 1: Classification of Carbohydrates

Functions

1. The primary function of carbohydrates in the body is to supply energy. Each gram of carbohydrate, as starch or sugar, provides

4 Kcal/g. It is known to exert a "protein sparing action". Due to the fact that in the presence of carbohydrates protein will not be oxidized by the body to yield energy.

2. Carbohydrates act as reserve energy supply in the form of glycogen, stored in muscles and liver. The total amount of glycogen in the body is over 300g.
3. Carbohydrate combine with nitrogen to synthesize non-essential amino acids.
4. Lactose, the milk sugar, provides galactose needed for brain development. It aids absorption of calcium and phosphorus, thus helping bone growth and maintenance.
5. Lactose forms lactic acid in the intestinal tract due to the action of the bacteria (lactobacilli). The presence of lactose promotes growth of lactobacilli. These lactobacilli synthesize some of the B-complex vitamins.
6. Carbohydrates are an important part of some compounds (Muco polysaccharides), which increase our resistance to infection.
7. Ribose, a five carbon sugar, is an essential part of DNA and RNA.
8. Dietary fiber helps smooth movement of food waste through the digestive tract and the soft, bulky stools are comfortably eliminated.

Starch is found in plant seeds (cereals and legumes), roots and tubers. Cereals which are the staple food in the Indian dietary, contribute major part of the energy and proteins, and some minerals and vitamins.

Table 3: Food sources of carbohydrates

Carbohydrates	Main Food Sources	Remarks
Polysaccharides		
Cellulose and hemicelluloses	Stalks and leaves of vegetables Outer covering of seeds	Indigestible
Pectin	Fruits	Indigestible
Gums and Mucilages	Plant secretions and seed exudates	Indigestible
Starch and dextrins	Grains, legumes and tubers	Digestible

Glycogen	Meats and seafood	Digestible
Disaccharides		
Sucrose	Cane and Beet Sugar, Molasses	Digestible
Lactose	Milk and Milk products	Digestible
Maltose	Malt products, some breakfast cereals	Digestible
Monosaccharides		
Glucose	Fruits, honey, corn syrup	Digestible
Fructose	Fruits, honey	Digestible
Galactose	Milk	Digestible

Legumes provide significant amount of proteins, iron and B-vitamins, in addition to energy. Potatoes, yams, jackfruit are good source of starch. Vegetables and fruits provide-dietary fiber, minerals and vitamins. Milk is the only animal food, which is the source of a very important sugar-lactose-in the diet.

Glucose the most common monosaccharide is present in honey, fruits and corn syrup. After absorption of food it is transported through blood and hence is present in blood as an easily available source of energy.

Sucrose is present in molasses, honey, fruits and vegetables. Sucrose is hydrolyzed to a mixture of equal amounts of glucose and fructose by action on digestive enzymes, when boiled with acid mixture of glucose and fructose is called as invert sugar.

Maltose is formed during sprouting of grains of digestion of starch by action of enzymes. It contains two glucose molecules.

Lactose is the sugar in milk. It yields glucose and galactose on hydrolysis by lactase.

Starch is found in cereal grains, legumes and in tubers.

Glycogen is found in meats & seafoods; these are digestible polysaccharides.

Indigestible polysaccharides–cellulose and hemicellulose, pectin, gum and mucilage's are indigestible polysaccharides. Cellulose are found in stalks and leaves of vegetables and outer coverings of seeds. Pectin is present in fruits and gums while mucilages are part of plant exudates and seeds.

Table 4: Carbohydrate content of foods

Food item	g/100g of edible portion of food	Foods	g/100g of edible portion of food
Sugar and Jaggery	85-99	Fruits	12-22
Rice flakes, Semolina, Vermicelli, etc.	75-78	Others vegetables	7-13
Wheat flour and Millets	69-72	Milk	4-5
Legumes and Pulses	58-60	Fish and seafood	1-4
Roots and Tubers	22-38	Meat and poultry	0-0.5
Nuts and Oilseeds	12-27	Fats and oils	0

RDA: The total energy intake of 65-70% should be from carbohydrate, remaining 19% from fat and 11-13% from protein.

Important Reactions in Carbohydrate

a) Browning Reaction

Browning is common colour change seen in food during pre-preparation, processing, or storage of food, it occurs in varying degrees in some food materials. The colours produced range from cream or pale yellow to dark brown or black, depending on the food item and the extent of the reaction. Browning reactions may be desirable or they may be undesirable. In some foods, the brown colour and flavour developed during browning is highly desirable and is associated with a delicious, highly acceptable, and quality product. Browning reactions contribute to the aroma, flavour, and colour of the product, such as the brown crust of bread and all baked and other processed foods.

Controlled browning is necessary even in foods where browning is desired because excessive browning can produce an undesirable product changes in odour and flavour, which may be bitter, making the food unpalatable.

Types of Browning Reactions

Browning reactions observed in food may be classified as enzymatic browning and non-enzymatic browning.

Non-enzymatic browning may be further classified as shown below:

Non-enzymatic browning

- Maillard reaction
- Caramelization
- Lipid browning
- Ascorbic Acid browning

b) Enzymatic browning

Some light-coloured fresh fruits and vegetables darken when exposed to air as a result of the presence of oxidative enzymes. Enzymatic browning occurs in these fruits and vegetables tissues when cellular organization is disrupted by cutting, bruising, or other injury to the tissues. This is due to the action of oxidative enzymes on the phenolic substances present in the fruit or vegetable tissues. Apples, bananas, pears, brinjals and potatoes undergo enzymatic browning. The cell contents come in contact with each other resulting in browning of uncooked fruits and vegetables which are cut or bruised.

Enzymatic browning takes place only in fruits and vegetables which contain phenolic compounds. These phenolic compounds act as the substrate, and in the presence of oxygen and by the action of enzymes, the following oxidative reaction is observed:

c) Caramelization

When sucrose (sugar) is heated above its melting point it undergoes a physical change to produce caramel. This happens more readily without water, however syrups will caramelize with rapid heating. This process is used extensively in the production of confectionary. Overheating will cause the substance to become bitter and dark.

d) Gelatinisation

When starch is mixed with water and heated, the starch granules swell and eventually rupture, absorbing liquid which thickens the mixture at temperature 60-88°C. Starch- a complex carbohydrate, has powerful thickening properties. When starch is combined with water or another liquid and heated, individual starch granules absorb the liquid and swell. This process, known as gelatinization, is what causes the liquid to thicken.

When exposed to heat, sugar will at first melt into a thick syrup. As the temperature continues to rise, the sugar syrup changes colour, from clear to light yellow to a progressively deepening brown. This browning process is called caramelization. It is complicated chemical reaction, and in addition to colour change, it also causes the flavour of the sugar to evolve and take on the rich complexity that we know to be characteristic of caramel.

Different types of sugar caramelize at different temperatures. Granulated white sugar melts at 320°F/160°C and begins to caramelize at 338°F/170°C. In foods that are not primarily sugar or starch, a different reaction, known as the Maillard reaction, is responsible for browning.

This reaction involves sugars and amino acids (the building blocks of protein). When heated, these components react and produce numerous chemical by-products, resolution in a brown colour and intense flavour and aroma. It is this reaction that gives coffee, chocolate, baked goods, dark beer, and roasted meats and nuts much of their rich flavour and colour. Though the Maillard reaction can happen at room temperature, both caramelization and the Maillard reaction typically require relatively high heat (above 300°F /149°C) to occur rapidly enough to make an appreciable difference in foods. Because water cannot be heated above 212°F/100°C unless it is under pressure, foods cooked with moist heat (boiling, steaming, poaching, stewing) will not brown.

Foods cooked using dry-heat methods (sautéing, grilling, or roasting) will brown. It is for this reason that many stewed and braised dishes begin with an initial browning of ingredients before liquid is added.

e) Retrogradation

Gelatinized starch is not in thermodynamic equilibrium. There is, therefore, a progressive re-association of the starch molecules upon ageing. This recrystallization is referred to as retrogradation, and may reduce the digestibility of the starch. The retrograding of the amylopectin component is long-term phenomena occurring gradually upon storage of starchy foods.

Prevention of enzymatic browning

Enzymatic browning can be prevented by the following methods:

By inactivating enzymes: Enzymes can be inactivated by any of the following measure:

Application of heat: Blanching or cooking fruits and vegetables which are prone to browning prevents discolouration. Enzymes are protein in nature and heat denatures proteins there by inactivating the enzymes.

Addition of salt: Vegetables may be immersed in a solution of sodium chloride to retard enzymatic browning. The chloride ion in NaCl inhibits enzyme activity. However, this is a temporary measure to retard browning, as the amount of salt required to prevent browning temporarily would make the food unpalatable.

Lowering the pH : Enzymatic browning is prevented by lowering the pH to 2.5 to 2.7 by addition of acid. Acids used to prevent browning are ascorbic acid, malic acid, citric acid, and lime juice. Ascorbic acid or vitamin C acts as an antioxidant and retards enzymatic browning. The quinones formed by the action of enzymes are reduced to their dihydroxyl state by vitamin C and in turn vitamin C is oxidized to dehydroascorbic acid.

Chilling a food below temperatures optimum for enzyme activity: Optimum temperature for enzymes to act is 43°C (109°F). In cold storage, the browning reaction slows down, but even fruits stored in frozen storage brown unless they are treated.

By avoiding contact with oxygen: Oxygen should not come in contact with the substrate. This can be achieved by any of the following measures:

Coating fruit with sugar: Coating fruit with sugar or covering it with syrup keeps atmospheric oxygen away from the surface. Intercellular oxygen is present in fruits, but sugar reduces the concentration of oxygen dissolved in the syrup and at the same time suppresses enzyme activity.

Immersing vegetables in water: Contact with atmospheric oxygen can be avoided by immersing cut vegetables in water. Since water contains dissolved oxygen, it is more effective if it is first boiled to remove dissolved air.

Vacuum packing: Protecting food from contact with oxygen as in vacuum packing prevents enzymatic browning since it is a oxidation reaction.

By elimination of substrate: The method of prevention is not practical. A variety of peach called 'sunbeam' is an exception because it is deficient in substrate.

Sulphuring of fruits prior to dehydration: Sulphur prevents oxidative browning due to enzyme activity. Fruit is treated to suphur fumes prior to drying. Treatment with sulphur dioxide (SO_2) gas or sulphurous acid solution (H_2SO_3) or 0.25 per cent sodium sulphite for 45 seconds is adequate to prevent browning. Bisulphites and metabisulphites are also used. Sulphurous acid is a strong reducing agent and prevents discolouration.

Non-enzymatic Browning

Maillard reaction

Maillard was the first to describe the development of a brown colour in mixtures containing amino acids and reducing sugars. The reaction between certain free groups of amino acids, such as the NH_2 amino group and a carbohydrate, affects the product in many ways in addition to the colour change. The aroma and flavour of the ready product is also affected. The reaction is also known as carbonylamine reaction or protein sugar reaction.

Process of Maillard reaction

Condensation of amino group of protein + Carbonyl group of sugar. Various reactions, such as rearrangement, fragmentation and polymerization.

Brown pigment

The brown pigment formed contributes to the aroma, flavour and colour of many ready-to-eat cereals, toffees, malted barley, and bakery products, such as bread, cakes, and biscuits. If the dough contains less reducing sugar, then the colour of the crust is light. The more the percentage of reducing sugar, the darker the crust in bakery products, such as dinner rolls.

f) Dietary fiber

Parts of plants (mainly carbohydrates), which are not digested by enzymes in the human intestinal tract are known as dietary fibers.

Structural parts of plant cell walls (cellulose, hemicellulose and lignin) which are not soluble in water, are classed as insoluble fiber. Water insoluble fibers are found in whole wheat, wheat bran, seeds, nuts and vegetables.

Viscous plant fibers are soluble in water. These include pectins, gums and mucilages. Soluble fibers are found in fruits, oats and barley, legumes and beans.

Functions and benefits of dietary fiber

i) They stimulate the peristalitic (rhythmic) movements of the gastrointestinal tract by adding bulk to the intestinal contents thereby helps to eliminate food waste from the body. Adequate fiber in the diet can prevent constipation. The presence of fiber increases gut motility. This helps in quicker evacuation of waste products of digestion from the body, thereby reducing risk of colon cancer.

ii) Presence of dietary fiber slows down carbohydrate absorption, which helps to regulate blood sugar and lowers insulin need of diabetics.

iii) Soluble fiber binds bile acids and cholesterol and helps carry these out of the body.

Limitations of fiber: Fiber binds some minerals, such as calcium, iron, zinc, etc.,

How much fiber is required: There is no recommended dietary allowance for fiber. Fiber intake can be upto 25-30 grams per day.

2. Proteins

Proteins are very large organic compounds, Proteins, like carbohydrates and fats, contain carbon, hydrogen and oxygen. Proteins contain about 16 per cent nitrogen, this distinguishes them from carbohydrates and fats. Some proteins also contain Sulphur, phosphorus, iron and cobalt.

Next to water, protein is the most abundant component of the body. It accounts for about a sixth of the live body weight and a third of it is found in the muscles, a fifty in the bones and cartilage, a tenth in the skin and the remainder in other tissues and body fluids. Proteins are present in all living tissues, both plant and animal. They are essential to life because vital parts of the nucleus and protoplasts of every cell are proteins.

Structure: All proteins in our body and food are built from basic units or compounds is known as amino acids. Amino acids combine to form proteins by means of a peptide bond, which joins a carboxylic carbon of one amino acid with nitrogen of another. The resulting peptide has a free carboxyl at one end and a free amino group at the other, permitting addition of other amino acids at either end.

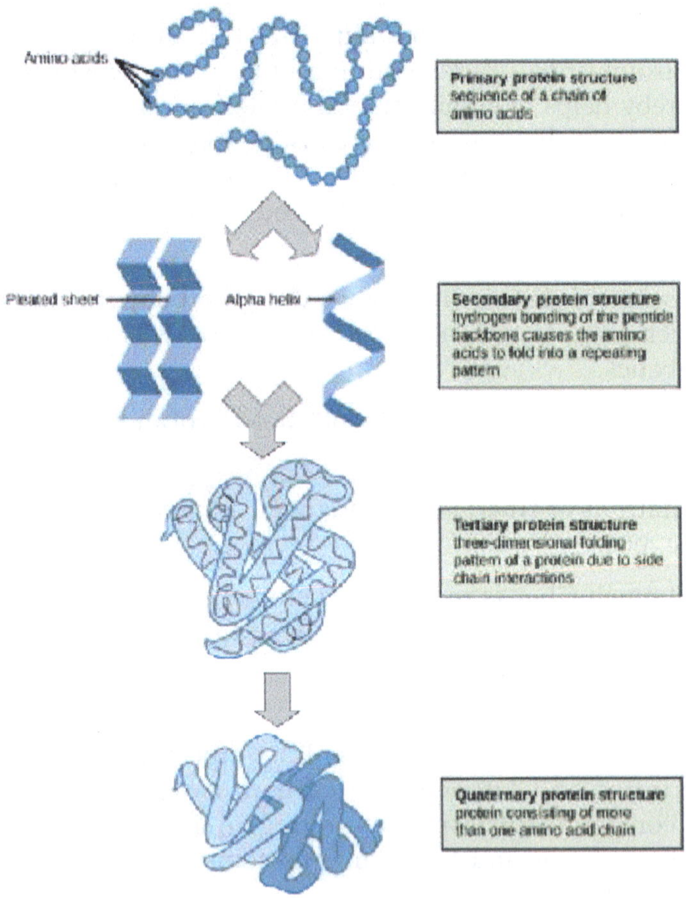

Fig. 2: Structure of protein

The proteins are **amphoteric** in nature due to the presence of a free acid and a free amino group in their molecule. **Polypeptides** thus formed constitute the **primary structure**. The **secondary structure** is formed by winding of the amino acid chain into helical or a pleated structure. Further folding, coiling and bending of polypeptide chains produces a three dimensional **tertiary structure**. Linking several polypeptide chain and in which the polypeptide chain is wound into a globular form results produces **quaternary structure**.

Proteins exist in **fibrous or globular** form. Examples of fibrous proteins are collagen of connective tissue, myosin of muscle tissue and keratin of hair. Globular proteins occur in tissue fluids; casein, egg albumin, albumins and globulins of blood plasma and haemoglobin. In conjugated form, they form most of the intracellular enzymes.

Classification

I. Classification based on structure and function

Proteins are broadly classified into two groups taking into account their structure and function:

Fibrous protein: Proteins are long and are less soluble in water. They are structural proteins, such as collagen, elastin, keratin and fibrin.

Globular protein: Proteins are compact and spherical, and are usually soluble in water. Enzymes, hormones, antibodies, transport proteins and respiratory proteins.

II. Nutritional classification

Proteins may also be classified according to their amino acid composition.

Complete proteins: Protein contains all the essential amino acids in adequate quantities This type of protein maintains the regulatory functions of body and promotes growth. *e.g.* proteins of milk, egg, meat.

In complete proteins: Proteins lack one or more essential amino acids. This type of protein can neither maintain the regulatory functions of body nor promote growth. *e.g.* gelatin.

The human body has certain limited powers of converting one amino acid into another. This is achieved in the liver by the process of trans amination, whereby an amino group is shifted from one molecule across to another under the influence of amino transferases, the co-amina acids are classified as essential amino acids.

Non-essential Amino Acids

Non-essential amino acids are ones that the body can make in adequate amount if nitrogen is available in the diet. They are non-essential only in the sense that they are not essential components of the diet.

Table 5: Classification of amino acids

Essential amino acids	Conditionally essential amino acids	Non-essential amino acids
Histidine	Arginine	Alanine
Isoleucine	Cysteine	Asparagine
Leucine	Glycine	Aspartic acid
Lysine	Proline	Glutamic acid
Methionine	Tyrosine	Glutamine
Phenylalanine		Serine
Threonine		
Tryptophan		
Valine		

Conditionally Essential Amino Acids

These are needed in the diet unless abundant amount of their precursors are available for their synthesis. The new born may not have enzymes in adequate amount to synthesise non-essential amino acid, or in intestinal metabolic dysfunction arginine may not be synthesized. Hence, it becomes conditionally essential amino acid. Amino nitrogen is not freely interchanged between all amino acids.

The precursors of conditionally essential amino acids are given below:

Table 6: Precursors of Conditional Essential Amino Acids

Amino Acids	Precursors
Cysteine	Methionine, serine
Tyrosine	Phenylalanine
Arginine	Glutamine/glutamate, aspartate
Proline	Glutamate
Glycine	Serine, Chloline

Partially complete: In this type of protein all the essential amino acids are present but one or more may be in lower quantities than required. Such a protein may be adequate to maintain the regulatory processes but cannot promote growth. e.g. proteins of cereal, pulses.

Functions

The functions of protein may be broadly divided into three:

1. Body-building or New Tissues
2. Regulatory
3. Energy supply

1. **Body-building or New Tissues**: Protein is an essential part of every cell amino acids must be supplied in the diet for building new tissues. Proteins provide the amino acids needed for the formation of new cells. They also provide the material from which nucleic acids are formed, e.g., DNA and RNA which carry the genetic code.

   The **primary function** of proteins is tissue building: Proteins are constituent of muscles, blood matrix of bones, teeth, skin, nails and hair. Our body, which consists of about 60 per cent water and 19 per cent fat, is held together by only 17 per cent proteins and 4 per cent minerals.

   The amount needed at various stages of life varies with the rate of growth is the highest during infancy. More protein is needed in the last part of pregnancy.

2. **Regulatory functions:**
   - Haemoglobin: A proteins and iron complex ensure the smooth running of respiratory cycle by being the vital oxygen carrier in the red blood cells.
   - Plasma proteins: contribute to osmotic pressure and thus, exert an important influence on the exchange of water between tissue cells and the surrounding fluids and on the water balance of the body as a whole.
   - Amphoteric nature of proteins is useful in maintaining acid base balance of blood and tissues.
   - Enzymes, Hormones and Antibodies : A small amount of protein is needed for synthesizing enzymes, hormones and

antibodies. All enzymes are proteins and are essential catalysts in digestion and metabolic processes in the tissues. Hormones, such as insulin, throxine, growth hormones, steroid hormones, etc., secreted by various glands are proteins in nature. Hormones regulate and co-ordinate body processes and activities. Antibodies and special white blood cells defend the body against infection and disease (immune system).

- Transport of nutrients: Due to their amphoteric nature, proteins are ideal carriers of nutrients across the cell membranes. Lipoproteins transport triglycerides, cholesterol, phospholipids and fat soluble vitamins across the cell wall.
- Special Functions of Amino Acids: Tryptophan serves as a precursor for the B-vitamin niacin and serotonin, a neuro-transmitter. Histidine is used in the synthesis of histamine used as a vasodilator in circulatory system.

3. **Energy supply function:** Each gram of food protein yields four calories to the body.

Food sources of protein

- Plants are the primary source of proteins in nature. Proteins re synthesized by the plants from the nitrates and ammonia in the soil.
- Pulses, whole and split, nuts and oilseeds, milk, eggs, fish, poultry and meat are good sources of protein in the diet.
- Cereals and their products are a major source of protein in the India dietary (Table 7).

Table 7: Protein content (g/100g edible portion) of common foods

Cereals	Protein	Animal foods	Protein
Wheat	12.8	Meat and Fish	15-20
Rice	7.5	Egg	13
Ragi	7.3	Milk	3.2-4.3
Pulses		**Vegetables**	
Dhals	22-24	Leafy vegetables	
Whole legumes	17-24	Tender beans/peas	4-9
Soya bean	43	Other vegetables	1-3
		Fruits	0.2-1.5

Protein Digestion, Absorption and Utilization

Protein digestion starts only in the stomach as there is no protein splitting enzymes in the saliva. In the stomach, protein is hydrolysed by the enzymes with the help of hydrochloric acid into peptides and amino acids. This process continues in both the duodenum and the small intestine, where the pancreatic and intestinal enzymes complete the hydrolysis by converting almost all the protein into amino acids.

Protein quality

There are 22 (or more) amino acids, all of which are vital to human life and health. The body is able to synthesis 13 of these but is unable to make 9 others. These nine amino acids have to be provided in the diet in sufficient amounts and are called essential amino acids - Histidine, tryptophan, valine, threonine, isoleucine, lysine, luecine, phenylalanine and methionine. Arginine is called semi-essential, because it may not be synthesized in sufficient amounts to meet the rapid growth of infants, especially premature one and may need to be supplied in the diet. Non-essential amino acids are termed non-essential is because these need not be provided in the diet.

Measurement of protein quality

There are several methods to determine the quality of protein in the diet. These are biological value (BV), net protein utilization (NPU), protein efficiency ratio (PER) and chemical score.

Biological value: Biological value is the percentage of the absorbed nitrogen (N) retained by the body.

$$BV = \frac{\text{N retained by the body}}{\text{N absorbed}} \times 100$$

The quality of protein is directly related to its BV. The BV of milk is 84, brown rice 73 and whole wheat 65. Due to complementary effect of food mixtures, the BV of mixed diets is higher than the arithmetical average of the component of food proteins.

Net Protein Utilization (NPU): Net protein utilization is digestibility of protein multiplied by its BV. NPU of Indian diet is 65.

Protein Efficiency Ratio (PER): In this method, a known amount of test protein in an adequate diet is fed to young rats for four weeks and weight gain determined. The PER is obtained by dividing the weight gain by grams of proteins fed.

$$\text{PER} = \frac{\text{Weight gain in g}}{\text{Protein fed in g}}$$

The PER of milk is reported to be 3.0 and polished rice 2.2

Complimentary (Supplementary) effect of foods to enhance the protein quality

The food proteins vary greatly in their amino acid composition. Protein quality depends on the kinds and amounts of the essential amino acids present in the food proteins. It is reported that certain food proteins lack one or more essential amino acids. For example, cereals are low in lysine; most of the pulses contain a small amount of methionine. Most of the animal proteins, such as milk, eggs poultry, fish and meat contain all essential amino acids. Cereals and pulses are normally consumed together with other foods, such as vegetables, curd, etc., the lack in one food is supplemented by the peptides or amino acids. This process continues in both the duodenum and small intestine. Where the pancreatic and intestinal enzymes complete the hydrolysis by converting almost all the protein into amino acids.

Denaturation of proteins

Denaturation of proteins involves the disruption and possible destruction of both the secondary and tertiary structures. Since denaturation reactions are not strong enough to break the peptide bonds, the primary structure (sequence of amino acids) remains the same after a denaturation process. Denaturation disrupts the normal alpha-helix and beta sheets in a protein and uncoils it into a random shape.

Denaturation occurs because the bonding interactions responsible for the secondary structure (hydrogen bonds to amides) and tertiary structure are disrupted. In tertiary structure, there are four types of bonding interactions between "side chains" including: hydrogen bonding, salt bridges, disulfide bonds, and non-polar hydrophobic interactions which may be disrupted. Therefore, a variety of reagents and conditions can cause denaturation. The most common observation in the denaturation process is the precipitation or coagulation of the protein as indicated in Fig. 3.

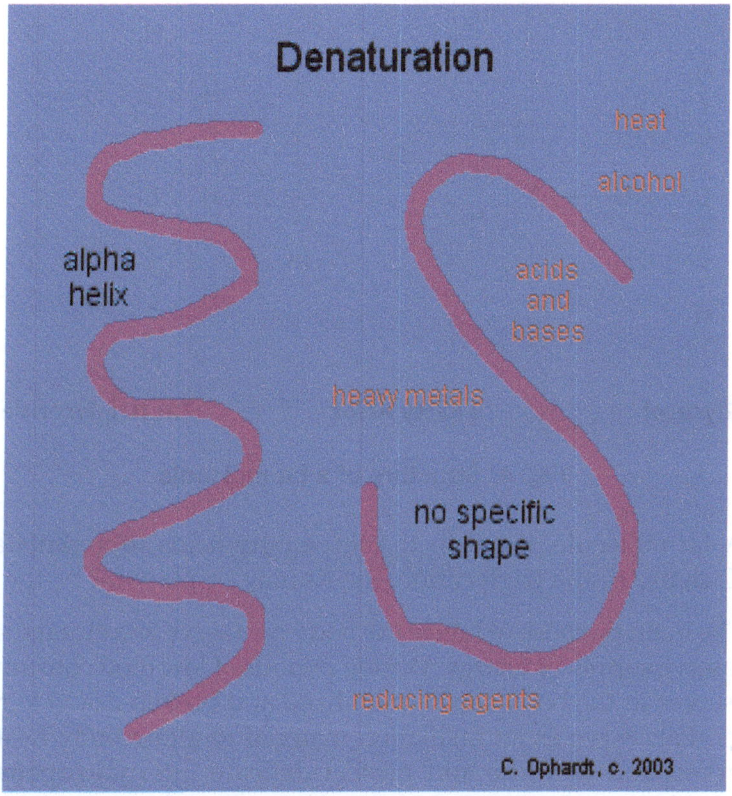

Fig. 3: Denaturation of proteins

Heat: Heat can be used to disrupt hydrogen bonds and non-polar hydrophobic interactions. This occurs because heat increases the kinetic energy and causes the molecules to vibrate so rapidly and violently that the bonds are disrupted. The proteins in eggs denature and coagulate during cooking. Other foods are cooked to denature the proteins to make it easier for enzymes to digest them. Medical supplies and instruments are sterilized by heating to denature proteins in bacteria and thus, destroy the bacteria.

3. Fats

Fats are a subgroup of compounds known as lipids that are found in the body and have the general property of being hydrophobic (meaning they are insoluble in water). Fats are also known as triglycerides, molecules made from the combination of one molecule of glycerol with three fatty acids (Fig. 4).

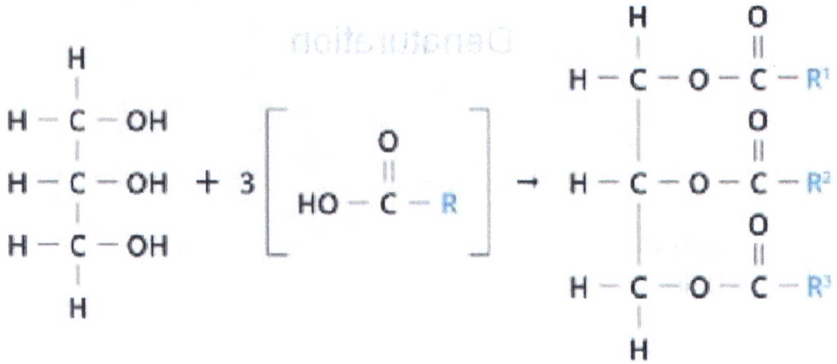

Glycerol | 3 Fatty Acids | Fat (triglyceride)

Fig. 4: Structure of a fat molecule

A fat molecule, the R in the three fatty acids represents a long C-C-C chain. In the triglyceride, the *Rs* may or may not be the same.

The main purpose of fats in the body is to serve as a storage system and reserve supply of energy. During periods of low food consumption, fat reserves in the body can be mobilized and broken down to release energy. Fats serve as an insulation material to allow body heat to be conserved and fats line and protect delicate internal organs from physical damage. Fats in the diet can be converted to other lipids that serve as the main structural material in the membranes surrounding our cells. Fats are also used in the manufacture of some steroids and hormones that help regulate proper growth and maintenance of tissue in the body. Lipids are organic compounds composed of a carbon skeleton with hydrogen and oxygen substitutions. Nitrogen, sulphur and phosphorous are also present in some lipids.

Properties

Every fat molecule has glycerol and one, two or three fatty acids attached to it. The nature of the fat or the oil depends on the kind of fatty acids linked to the glycerol. The hardness, melting point and the flavour of the fat is related to the length of the carbon chain and the level of the saturation of the fatty acid. If the substance is a liquid at 20°C, it is called oil, if solid at that temperature, it is known as fat. Saturated fatty acids are found in solid fats, whereas most of the oils contain unsaturated fatty acids.

Classification of fats

Lipids are classified into three groups on the basis of their chemical structure:

1. Simple lipids, which include fats and oils. (triode and mono glycerides)
2. Compound lipids, which comprise of phospholipids and lipoproteins, and
3. Derived lipids, such as fatty acids and sterols.

   - **Glycerides**: Fatty acids combine with glycerol to form a glyceride. When only one fatty acid combines with glycerol, it forms a monoglyceride. Diglycerides have two fatty acids and triglycerides three fatty acids attached to glycerol. Most of the fatty acids in the body as also in foods occur in the form of triglycerides.
   - **Phospholipids:** Phospholipids are important as components of active tissues (brain, nervous tissue and liver). They are powerful emulsifying agents and are essential for the digestion and absorption of fats. Phospholipids help to carry lipid particles across the cell membrane in the blood stream, e.g. lecithin, sphingomyelins and cerebrosides.
   - **Sterols**: Three sterols with functions associated with nutrition are:

     Ergosterol a plant sterol.

     Cholesterol an animal sterol.

     Cholesterol is an important sterol. It is a precursor of vitamin D, hormones and bile acids.

     Ergosterol and 7-dehydrocholesterol are two precursors of vitamin D.

Lipoproteins: Lipoproteins are synthesized in the liver. They are composed of three-fourth to two-third lipids and the remainder protein. These serve as the major vehicle for fat transport in the blood stream.

Four groups of lipoproteins that have been identified are chylomicrons, low density lipoproteins (LDL), high density lipoproteins (HDL), and very low density lipoproteins (VDL). The LDLs carry fat and cholesterol to cells. The HDLs carry free cholesterol from body tissues to the liver for breakdown and excretion.

Fatty acids

Fatty acids are the main building blocks of fats. Fatty acids have a methyl group(CH3) at one end and a carboxyl group (COOH) at the other end with a chain of carbon and hydrogen atoms in the middle. The fatty acids have three important aspects, chain length, saturation and essential fatty acids.

The number of carbon atoms in fatty acid decides the chain length, thus short chain fatty acids contain 4 to 6 carbon. The number of carbon atoms, long chain fatty acids have 14-18 carbon atoms and extra-long chain fatty acids have more than 20 carbon atoms in the molecule, milk, fat, vinegar and coconut contain short chain fatty acids, long chain fatty acids are found in most vegetable oils and animal fats. Fish oils contain the extra-long chain fatty acids.

Saturated fatty acid has a single bond between its carbon atoms, thus the molecule is saturated with hydrogen. Stearic acid is an example of a saturated fatty acid. Unsaturated fatty acids have one or more double bonds in their molecule and are thus not saturated with hydrogen.

Mono saturated fatty acids (MUFA) have only one double bond in the molecules oleic acid, found in olive oils, is a monounsaturated fatty acid.

Polyunsaturated fatty acids (PUFA) have two or more double bonds in the molecules. Corn oil contains polyunsaturated fatty acids.

[The methyl end of the fatty acid molecule is called the omega end. If the first double bond in the fatty acid is in the third carbon atom from the methyl end, it is called omega-3 fatty acid; if it is in the sixth carbon from the methyl end, it is called omega-6-fatty acid and when it is called omega-6-fatty acid and when it is in the ninth carbon, it is called omega-9 fatty acid.]

Examples

Oleic acid-Omega-9-fatty acid, which is a MUFA

Linoleic acid, omega-6 fatty acid, which is a PUFA

Linolenic acid, omega-3-fatty acid, which is also a PUFA

Oleic (18C) and linoleic (18C) are some of the common unsaturated fatty acids.

Hydrogenated fats: The unsaturated fatty acids can be converted to saturated one, thus changing the physical and chemical characteristics of the oil, by a process known as hydrogenation in which hydrogen is added on to the oils, in the presence of a catalyst.

Trans fatty acids: Fatty acids are usually present in foods naturally in cis configuration. When the conversion from cis to trans configuration occurs while processing e.g. hydrogenation then trans fatty acids are produced and these are harmful to human health.

Essential fatty acids (EFAs) : Linoleic acid (omega-6 PUFA) and linoleic acid (omega-3 PUFA) are called essential fatty acids because:

1. These are not-synthesized in the human body.
2. These are required for important functions in the body. They are essential for growth in the young and maintenance of normal healthy skin. They are precursors of eicosanoids, a group of important metabolites which regulate vascular function, one of these are prostaglandins.
3. These are available only through diet.

Dietary sources of EFA : Common plant oils with the exception of coconut are good sources of linoleic acid. Green leafy vegetables, cabbage and lettuce are good sources of omegar-3-fatty acids. Soyabean and grape seed are also good sources of omega-3-fatty acids. Fatty fish, such as tuna, sardines and salmons are good sources of omega-3-fatty acids.

Functions of fat

1. Oils and fats are concentrated sources of energy. Each form of fat supplies 9 calories of energy.
2. Fat acts as a storage form of energy. It is stored in various parts of the body as adipose tissue.
3. Food fats are the source of essential fatty acids(EFA)
4. Food fats also aid the transport and absorption of fat-soluble vitamins.

5. Cholesterol is an essential lipid. Some important hormones and bile acids are formed from cholesterol.
6. Fats are used to synthesis phospholipids, which are found in all cases.
7. Fats act as a cushion for certain vital organs.
8. Since fat is a poor conductor of heat, a layer of fat beneath the skin helps to conserve body heat and regulate body temperature.
9. The flavour, palatability and satiety value of foods increases by fats. The slower rate of digestion of fat as compared to carbohydrates results in a feeling of satisfaction.

Digestion and absorption

- The digestion of fats starts in the stomach. Most of the fat digestion occurs in the small intestine. The entry of fat in the duodenum, the first part of the small intestine stimulates secretion of the bile from the gall bladder. The fat gets emulsified by the bile preparing it for digestion.
- The fat is broken into small particles, thus increasing the surface area and the surface tension is lowered improving the enzyme (pancreatic lipase) action.

Table 8: Sources of fats

Food	Fat g/100g of food	Calories/100 g
Visible fats		
Oil, vegetables	100.0	900
Vanaspati	100.0	900
Ghee, cow's	99.5	895
Butter	81.0	729
Invisible fats		
Oilseeds and nuts	37.0 - 64.5	537 - 681
Mutton, muscle	13.3	194
hen	13.3	173
Liver, sheep	7.5	150
Milk, cow	4.1	67

The products of digestion are absorbed through the walls of the small intestine and circulated through the lymph. Some of these are used to synthesize important lipid compounds. Some fat is used to supply energy. The rest is stored as fat in the adipose tissues.

Requirement: The fat requirement is based on two factors, namely to meet energy and the essential fatty acids needs. About 10 per cent of the total energy need is met by invisible fat in the diet.

A minimum of 5 and maximum of 10 per cent of total energy needs to be provided as visible fat in the diet. This works out to about, 12g -20g of fat per day. A higher, level of intake of 20g/day is desirable to provide energy density and palatability for normal adults.

Disorders of fat intake

Energy intake in excess of needs lead to obesity and an unacceptable blood lipid profile. There is deposition of fatty material with formation of plaques in the arteries, leads to atherosclerosis, high blood pressure.

- Animal fat, a source of saturated fats in the diet, are also high in cholesterol (CHOL) content. The saturated fats are known to increase the total CHOL and LDL production in the liver, leading to increased levels in the liver, leading to increased levels in the blood plasma. Increased serum CHOL level above 200mg/dl and low density lipoproteins (LDL) above 130 mg/dl are associated with increased risk of coronary heart disease(CHD). The high-density lipoproteins (HDL) level in the blood is a beneficial factor, as it protects the blood vessel against formation of fatty deposit (atheroma), by moving the fatty material away.

4. Energy

Energy is defined as the ability to do work. Next to air and water, the body requires food as a continuous source of energy to stay alive and keep all organs and systems functioning efficiently. Just as every engine requires fuel to keep going, the human body too requires fuel in the form of food as a source of power to work continuously. The human body is far more complex than any machine invented and as long as there is life, its energy supply cannot be turned off.

The first and foremost function of food is to supply energy to the body. This takes priority over building of new tissues, repair of wear and tear, and regulation of body functions. For example, if a diet contains adequate protein but is deficient in carbohydrate and/or fat, the protein will be oxidized to meet the energy needs first and balance used for other functions.

When food is digested, the complex nutrients carbohydrates, fats, and proteins are broken down into monosaccharides, fatty acids, glycerol, and amino acids, respectively. These simple forms are absorbed into the bloodstream and supplied to the millions of cells in the body to be oxidized by a series of complex steps to release energy.

Glucose

Fatty acids + Oxygen CO_2 + H_2O + energy

Glycerol

Amino acids deamination ($-NH_2$)

(In sugar or fat is deficient)

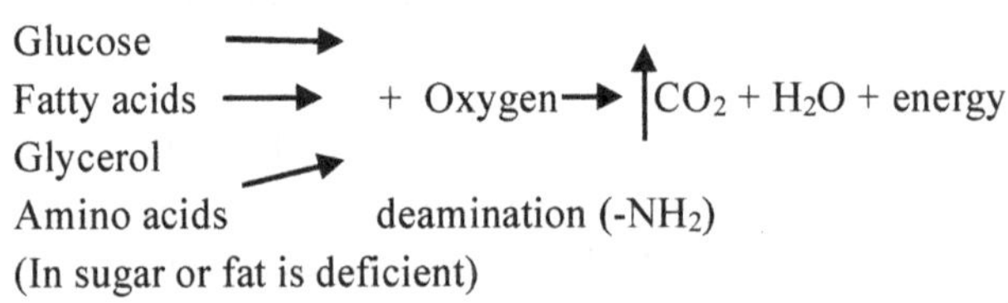

Fig. 5: Oxidation of nutrients to release energy

Forms of Energy

Energy exists in several forms. The forms of energy important in nutrition are :

1. Chemical energy in food.
2. Light or solar energy for synthesis of vitamin D in the skin and for photosynthesis in plants.
3. Mechanical energy for movement of muscles.
4. Electrical energy for functioning of the brain and nerve cells.
5. Heat energy, generally produced when energy is converted from food is finally converted into heat energy.

The various forms of energy are interconvertible

Living cells are capable of releasing the energy stored in certain nutrients. This energy is used to perform various activities in the cell, such as synthesis of proteins, maintaining warmth, and contraction of muscle. The energy from the breakdown of food is stored in the body in the form of a high energy compound, adenosine triphosphate (ATP). Adenosine triphosphate acts as a store of energy -rich phosphate bonds. Living cells can use energy only in the form of energy-rich phosphate bonds.

When energy is required for cellular activity, converted to adenosine diphosphate (ADP). One mole (molecular weight in gram) of ATP provides 8 kcal/33 kJ of energy. Adenosine diphosphate is reconverted to ATP by addition of phosphate ions. Phosphate ions help in release and transfer of energy.

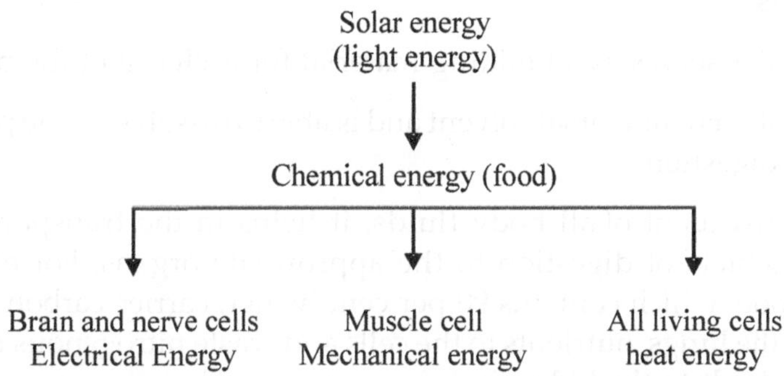

Fig. 6: Various forms of energy are inter convertible

Units of measurement

The energy present in food or the energy needed by the body is measured in units called joules or calories. The calorie in nutrition is the large calorie or kilocalorie. The kilocalorie is defined as the amount of heat required to raise the temperature of 1 kg (1,000 g) of water by 1°C. This calorie is 1000 times bigger than the calorie used in physics.

JOULE (J) : The international unit for energy is the joule (J) and it is the energy expended when 1 kilogram (kg) is moved 1 metre (m) by a force of 1 newton (N).

Scientists and nutritionists are concerned with large amounts of energy, so they use the units kilocalorie, kilojoule (kJ = 10^3 J), or megajoule (MJ= 10^6 J) to express energy.

1 kilocalorie = 4.184 kilojoules

1 megajoules = 239 kilocalories

5. Water

Water is the major component of our body. Approximately 55 to 70 per cent of the total body weight is made up of water. The percentage of water tends to decrease as a person gets older. Fat individuals have less water than lean ones. The fluids, which exist inside the cells, are called intracellular fluids, which form about 55 per cent of water in the body. The rest is found as extracellular fluid (outside the cells). Blood, lymph circulation and interstitial fluid (fluid between cells or tissues) are part of extracellular fluid. Kidneys are the final regulators of fluid balance.

Functions

1. Water serves as a building material for each cell of the body.
2. Water is a universal solvent and is able to dissolve all the products of digestion.
3. Constituent of all body fluids, it helps in the transport of the products of digestion to the appropriate organs. For example, blood which contains 90 per cent 'water, carries carbon dioxide to the lungs, nutrients to the cells and waste nitrogenous material and salt to the kidneys.
4. Urine which contains 97 per cent water has all the waste material dissolved in it and the body is thus able to excrete soluble waste products of metabolism.
5. Water is needed for many chemical reactions to occur in the blood for example, the breakdown of sugar to simpler substances needs the presence of water.
6. Water acts as a lubricant preventing friction between moving parts of the body.
7. The body temperature is regulated through the evaporation of water from the skin and lungs.

Water balance : The body normally maintains a water balance the amount of water ingested is equal to the water excreted or lost from the body but certain regions of hypothalamus regulate the intake.

Table 9: Water balance in healthy adult

Water intake (ml/day)		Water loss (ml/day)	
Water intake	800 to 1100	Urine	800 to 1000
Water in food	600 to 900	Vapour (skin and lungs)	600 to 1000
Metabolic water	200	Faeces	200
Total	**1600 to 2200**		**1600 to 2200**

Sources: The water we drink intake of all beverages and liquid foods release water

The water excretion is controlled by hormones. Kidneys and lungs carry out water from the body. Water is also lost as perspiration through the skin and as part of excreta from bowel.

Requirement: Adults need, about 1 ml of water is needed per 1 kcal energy intake; thus about 2000 ml water is necessary when energy intake is 2000kcal. Infants who have a large body surface area, in proportion to body weight, need 1.5ml water /1 kcal energy intake.

The amount of water needed by an individual will depend on many factors, such as the environmental temperature, humidity, occupation and the diet.

Water is the major component of our body. Approximately 55 to 70 per cent of the total body weight is made up of water. The percentage of water tends to decrease as age advances.

Extracellular fluid: Per cent of body weight :Average subjects-20%, Fat subjects -15%, Thin subjects -25%.

Table 10: Distribution of water in the body

	As per cent of body weight
Total water	**70**
i) Intracellular	50
ii) Extracelluar	20
a) Plasma	4
b) Lymph	7
Interstitial water (tissue fluid)	9

6. Vitamins

Vitamins are one of the six classes of nutrients supplied by food. They are required for normal growth and maintenance of all animal life. Vitamins are important for their regulatory and protective functions. Unlike most other nutrients they are required in very small amounts. But it is necessary to provide these in the diet because many of them cannot be manufactured by the body. The lack of vitamins results in definite deficiency disorders, which are specific for each particular vitamin.

Definition: Vitamins are organic substances, which occur in small amounts in foods.

General functions

- Vitamins do not provide calories, but are essential in the metabolic reactions, which release energy from carbohydrates, fats and proteins.
- Vitamins are essential co-factors in metabolic reactions in the body.
- Vitamins may occur in preformed or its active form in the food, or as a precursor compound which changes into active form in the body.

Classification: Vitamins are classified into two groups on the basis of their solubility into fat-soluble and water-soluble vitamins.

Fat soluble vitamins include A, D, E and K

- Fat soluble vitamins can only be absorbed in the presence of fat. Therefore, the presence of some fat in the diet is essential for their absorption.
- Fat soluble vitamins can be stored in the body. The requirement for fat-soluble vitamins may be met by intake of precursor of the vitamin itself.
- Not much of fat-soluble vitamins are lost in normal cooking procedures.

Water soluble vitamins include the B-group and vitamin C.

- Water soluble vitamins can only be absorbed easily.
- Water soluble vitamins cannot be stored in the body.

- Appreciable amounts of water soluble vitamins are lost in normal cooking procedures.

Fat Soluble Vitamins

Vitamin A

This was the first fat soluble vitamin to be discovered. It has a number of important functions in the body. Vitamin A is found only in animal food, mainly as retinol. Plants provide a source of vitamin A for animals in the form of orange-yellow pigments called the carotenoids. The chief source in human nutrition is beta-carotene, which the body converts to vitamin A in the intestinal mucosa during absorption. The conversion of carotene to vitamin A is partial and varies from 25 to 50 per cent . Vitamin A activity in food used to be expressed in International Units (I.U) 1 IU=0.3 mcg retinol. Now it is expressed as a microgram (mcg) of retinol or beta-carotene or retinol equivalent (RE).

Functions

Growth: Necessary for formal growth and development.

Function in vision: Vitamin A occurs in the retina of the eye and is required in the process of vision to adjust to light of varying intensity (dark adaptation). Night blindness occurs in severe vitamin A deficiency; it indicates the inability of a person to see at night.

Health of epithelial tissues: These tissues cover the outer surface of the body. Inadequate supplies of vitamin A results in suppression of the normal secretions and produces a keratinized (dry, horny) type of epithelium. The skin may become excessively dry and mucous membrane may fail to secrete normally and hence be prone to bacterial invasion.

In vitamin A deficiency keratin -producing cells replace mucus-secreting cells in many epithelial tissues of the body. This leads to the drying of the conjunctiva and cornea of the eye. The process can be reversed by an intake of vitamin A.

Immune response: Many of the epithelial tissues are important barriers to infection. Vitamin A deficiency impairs this function.

Haemopoesis: Vitamin A deficiency is associated with iron deficiency anaemia.

Food sources

Vitamin A is present in animal foods only. Liver is the richest source of vitamin A. Other sources, include butter, ghee, milk, curds and egg yolk.

Refined oils and vanaspati are good sources of vitamin A. These may be fortified with Vitamin A to the extent of 750mcg (2500 IU) per 100g.

Vitamin A is not present in vegetable foods, but these foods contain the pigment, beta-carotene, which is the molecules of vitamin A. Leafy vegetables, such as spinach, amaranth, coriander, and drumstick leaves as well as ripe fruits such as mango, papaya, and yellow pumpkin are good sources of beta-carotene. Generally, dark green leafy vegetables contain greater amounts of carotene than those which are light in colour.

Large intake of vitamin A in times of plenty is stored in the liver and helps the body to tide over periods of shortage. Similarly, an occasional intake of a concentrated source also helps to meet the requirements for this vitamin.

Table 11: Sources of Vitamin A and β-carotene

Vitamin A foods
Liver, sheep
Eggs
Butter, ghee (cow's)
Milk
β-carotene foods
Green leafy vegetables-amaranth, spinach, cabbage, etc.
Yellow or orange, vegetables and fruits, e.g., carrots, mango, papaya, orange
Immature beans and peas, capsicum, tomato

Lycopene, a carotenoid from tomatoes, is an antioxidant. It has the ability to protect against vascular degeneration, the leading cause of adult blindness.

Requirements for vitamin A is expressed in terms of retinol equivalents (RE/day). The daily requirement of an adult for vitamin

A is of the order of 600 mcg of retinol or 4800mcg of beta-carotene per day derived from foods of either animal or vegetable origin.

Table 12: Recommended dietary allowances of vitamin A (mcg/day)

Age	Retinol Equivalents (RE)	Beta-carotene
0-6 months	350	2800
Infants (6-12 months)	350	2800
1-6 years	400	3200
7-12 years	800	4800
13-18 years	600	4800
Adult man and woman	600	4800
Pregnant woman	800	6400
Lactating woman	950	7600

Effect of deficiency

The deficiency may result from dietary lack of vitamin A, the provitamin, or poor absorption of these. The deficiency results in growth failure, affects the vision, the skin and the immune function adversely.

The earliest symptom is impaired ability to see in dim light (dark adaptation): the next stage is inability to see normally in dim light which is known as night blindness or nyctalopia. A later, and more severe stage of deficiency is xerosis (dryness) of the cornea and conjunctiva. The cornea becomes dry and loses its transparency, Keratomalacia. Cornea becomes soft and results in permanent blindness (Xerophthalmia).

In early stages, treatment with vitamin A will restore full vision: the skin changes include dryness (hyperkeratosis)Progressive degeneration of epithelial cells, which increases susceptibility to severe infections of the eye, the nasal passages, the sinuses, middle ear, lungs and genitourinary tract.

Vitamin D

Vitamin D is called the 'sunshine vitamin' because the body is able to convert a precursor 7-dehydrocholesterol, a sterol present in the skin to Vitamin D in the presence of sunlight. It can be synthesized in the body in adequate amounts by simple exposure to sunlight even for five minutes per day.

Functions

Absorption of calcium and phosphorus: Without the presence of vitamin D, formation of strong and rigid bones is not possible.

Bone mineralization: It regulates the rate of deposit and resorption of these minerals in bone. This balancing process helps to build and maintain bone tissue. Vitamin D can be used to treat rickets in children and osteoporosis (bone loss) in older women.

Food sources: Irradiation of the skin with sunlight is the main source of vitamin D. Foods are not a good source of vitamin D. The richest source known is fish liver oils. Fish liver oils do not form part of the diet and have to be taken as a supplement.

Recommended dietary allowance

About 5 mcg or 200 international units are stated to be the daily requirement of a child. However, during pregnancy, adequate supply of vitamin D is essential for the healthy development of the foetus.

Effect of deficiency

When adequate quantities of vitamin D are not available, strong and rigid bones are not formed. This leads to a condition in children known as 'rickets' which is characterized by poor growth and bone deformation, such as bowed legs, beaded ribs, enlarged joints and skull deformation. The teeth of children may fail to develop normally because of poor calcification.

Oesteomalacia- a condition similar to rickets may develop in adult women due to lack of vitamin. This condition is characterized by softening of bones resulting deformity.

Vitamin E

Vitamin E or alpha-tocopherol is a fat-soluble vitamin.

Function: Antioxidant: Vitamin E is a powerful fat-soluble antioxidant. Selenium is a trace mineral that works with vitamin E as an anti-oxidant.

Food sources of vitamin E are the vegetable oils. Vegetable oils are also the richest sources of polyunsaturated fatty acids, which vitamin E protects. Other food sources include cereals, leafy vegetable, milk, eggs, muscle meats and fish. Vitamin E is widely distributed in foods. Even the cheapest kind of cereal diet contains sufficient intake of the vitamin.

Vitamin E suggested is 0.8 mg/g of essential fatty acids.

Vitamin K

Phylloquinone is the major form of vitamin K found in plants. It is also found in our diets.

Functions

Blood - Clotting process: Vitamin K plays an important role as it is -

- Essential for the formation of prothrombin by the liver.
- Prothrombin is a normal constituent of the blood and helps clotting of blood in contact with air.

Food sources

- Green leafy vegetables are the principal dietary source of vitamin K. Human babies do not have the reserves of vitamin K at birth; therefore, routine practice is to give vitamin K are synthesized by the intestinal micro flora.
- Significant quantities of vitamin K are synthesized by the intestinal micro flora.

Deficiency: Deficiency of vitamin K prolongs clotting time and may result in excessive bleeding. Vitamin K deficiency is seen in India only occasionally in premature newborn infants. It is suggested that a dose of 0.5- 1.0 mg of vitamin K be administered by the intramuscular route to such infants. The normal formation of vitamin in the intestinal tract is disturbed by the use of sulfa drugs. Patients, who are treated with antibiotics, after surgery, are likely to suffer from vitamin K deficiency, due to killing of intestinal bacteria, which synthesize the vitamin. Thus, there may be blood loss due to non-clotting and poor wound healing.

RDA 55 mcg/day

Water soluble vitamins

Water soluble vitamins include ascorbic acid and the B-complex vitamins. The water soluble vitamins are absorbed quickly in the body and the amount not utilized are excreted in the urine. Adequate amount should, therefore, be supplied in the daily diet. Some of the water-soluble vitamins are partly lost in cooking procedures.

Thiamin, riboflavin, niacin, pyridoxine, folic acid and vitamin B_{12} are included in the RDA, because definite requirements of these vitamins have been established. A diet, which provide adequate amount of these six vitamins, also, carriers enough of the other members of this group. All these vitamins are essential for human nutrition.

Thiamin

Thiamin is known as vitamin B_1. It exists mainly as thiamin pyrophosphate (TPP), a coenzyme of vital importance in intermediary metabolism.

Functions

- The basic function of thiamin as a coenzyme is related to release of energy from glucose and its storage as fat. Thus, it makes energy available for normal growth and function of the body.
- Thiamin pyrophosphate, the coenzyme form of thiamin, is necessary for catalyzing the oxidation of carbohydrates in the body.

Food sources

- Plant sources include pulses, nuts, oilseeds and whole grain cereals. Cereal bran and germ are very rich sources of thiamin while refined flour is devoid of the vitamin.
- Parboiled rice and fresh peas are good sources of thiamin.
- Green leafy vegetables and animal foods, such as milk, eggs, fish and meat are fair sources.
- Organ meats, pork, liver and eggs also supply thiamin.
- Nearly all foods, except sugars, fats and oils, contain some thiamin. Most foods however, contain only small quantities and it is necessary to consume variety of foods to fulfill the daily thiamin needs.

Effect of processing : The main source of thiamin in the India dietary is cereals. Refining of cereals reduces thiamin content with most of the thiamin being lost in the polishing as thiamin is in the aleurone layer of the cereal grain. Thiamin is lost, when wheat flour is refined to manufacture maids. Using whole grain flours to make chapattis, rotis, ensures retention of thiamin. Parboiling of rice helps to conserve thiamin. Cooking the food to just adequate and avoiding use of soda in preparation, helps to retain most of the thiamin in the preparation.

Thiamin being water soluble is also lost-

- When water in which cereals or pulses are cooked is discarded.
- Long cooking also destroys some thiamin due to leaching.

Refinement and polishing of grains also loses thiamine. Thiamine is directly involved in energy and carbohydrate metabolism, the thiamin requirement is expressed in terms of energy intake. Hence, the recommended thiamin allowance is 0.5 mg per 1000 calories for all age groups.

Deficiency

Causes

- High carbohydrate diets, heavy alcohol intake aggrivate thiamine deficiency, due to low food intake and increased demand.
- Poor absorption due to abnormality of GI tract.

Symptoms of thiamine deficiency

Mild deficiency

- Loss of appetite or anorexia due to disturbed digestive system
- Decreased muscle tone
- Mental depression and confusion
- Neurological changes

Acute deficiency: It occurs in infants between 2 and 5 months of age.

Wet beriberi: Caused by eating polished, highly milled rice. Oedema is the most notable feature, affects the heart and hence, also known as cardiac beriberi.

Dry beriberi-peripheral neuropathy: Seen in alcoholics, wasting of muscle and tachycardia are present, affects the nerves (polyneuritis).

Prevention : Thiamin deficiency can be prevented by including foods rich in thiamin.

Riboflavin

Riboflavin is also called as vitamin B_2. It contains sugar 'ribose', hence the name riboflavin. Riboflavin is destroyed on exposure to sunlight.

Functions

- Riboflavin is an integral part of the coenzymes Flavin Mononucleotide (FMN) and Flavin- Adenine Dinucleotide (FAD).
- Functions as a vital part of coenzymes in both energy production and tissue protein building.

Deficiency : Deficiency of riboflavin results in a group of symptoms collectively the syndrome is called riboflavinosis.

- Lack of riboflavin affects the eyes, skin and nerves.
- Angular stomatitis, glossitis, cheilosis are characteristic features of riboflavinosis.
- Lesions at the angles of mouth- angular stomatitis.
- Inflammation of the tongue- glossitis.
- Dry appearance of the lips with superficial ulcers terms as cheilosis.
- Eye becomes sensitive to bright sunlight. This condition is called photophobia.

Food sources: Milk (including skim milk, butter milk, curds, cheese and whey), eggs, liver and green leafy vegetables are good source of riboflavin, whole grains are fair source, refined cereals are poor source.

Niacin

Niacin, also called nicotinic acid and nicotinamide is a vitamin of the B-complex group called vitamin B_3.

Functions: Niacin functions in the body as a :

- Component of two important co-enzymes **NAD** (Nicotinamide adenine dinucleotide) and **NADP** (Nicotinamide adenine dinucleotide and its phosphate).
- These coenzymes are involved in tissue respiration and synthesis and the breakdown of glucose to produce energy.

Sources

- Cereals, which are used as staples, are the major source of niacin in the India diet.
- Groundnuts are the best source of niacin.
- Liver, an organ meat, is a rich source of niacin.
- Unrefined and parboiled cereals retain more niacin than refined ones.
- Brewer yeast is a very concentrated source of niacin.

Tryptophan, an essential amino acid is precursor of niacin in the body. The total niacin requirement is stated in terms of 'niacin equivalents'. About **60 mg of tryptophan** can give rise to **one mg of niacin**.

Niacin equivalents (mg) = 60 mg of Tryptophan $\rightarrow$ 1 mg of Niacin

Recommended Dietary Allowance (RDA) 6.6 mg per 1000 K cal for all ages.

Effect of deficiency: Deficiency is seen predominantly among poor people whose staple is maize (pellagra is called as maize eaters disease). This is because maize is deficient in both niacin and also tryptophan. Among poor population whose staple food is jowar, the disease pellagra occurs due to an imbalance of amino acids.

Lack of niacin affects the skin, gastrointestinal tract and nervous system. The deficiency disease is known as pellagra. Pellagra has been called the disease of the three Ds."

- **Dermatitis** - affects the skin. Characterized by bilateral skin lesions. Skin exposed to sun is most affected - hands,

  Face and feet.
- **Diarrhoea** - Soreness of tongue and inflammation of FIT resulting in diarrohea.
- **Dementia** - affects the nervous system, disturbed mental state.
- Leading to death', the fourth D.

Treatment: Pellagra can be cured by a good diet containing adequate amounts of protein, tryptophan, niacin as well as other members of B-complex group of vitamins. Although milk is poor in nicotinic acid content it is effective in preventing pellagra because it is rich in essential amino acid tryptophan.

Folic acid

Folic acid is pteroyl -mono-glutamic acid.

Functions: The primary function of folic acid is related to the

- Transfer of single carbon in the synthesis of a number of metabolites in the body.
- Involved in the synthesis of nucleic acid along with vitamin B_{12}.

Sources : Green leafy vegetables, liver, legumes and yeast are rich sources of folic acid.

RDA 200 mcg per day for adults, during pregnancy is 500 mcg of additional folic acid daily. Nursing mothers need additional 300 mcg of folic acid.

Deficiency: leads to immature RBCs formation of red blood cells resulting in macrocyticmegaloblastic immature RBC anemia.

Vitamin B_{12}

Vitamin B_{12} or cyanocobalamin: An intrinsic factor in stomach helps to absorb the vitamin.

Functions

- Promotes normal growth and development .
- Component of several coenzymes for synthesis of nucleic acids.

Sources

The richest sources are liver, and organ meats. Animal foods are good sources. Plant do not contain vitamin B_{12}.

Deficiency: Inability of new RBCs to develop leading **to** pernicious or megaloblastic anaemia.

RDA - 1 mcg for men and women.

Ascorbic acid (Vitamin C)

The chemical name for vitamin C is ascorbic acid also known as hexuronic acid and antiscorbutic nutrient.

Functions

1. Vitamin C is a biological reducing agent especially during hydroxylation reactions.
2. Being a strong reducing agent, it helps to tie up free radicals and thus, protects the body against oxidative stress. As an antioxidant it protects the body against damaging oxidizing agents.
3. It keeps the iron in the ferrous state which is the form in which it is used by our body.
4. It is needed for Collagen formation. Collagen is a major structural protein of connective tissue (which binds cells and tissues together), bone, teeth, cartilage and skin.
5. It is needed in wound healing.
6. It helps the body to build resistance to infection.
7. It helps in the absorption of calcium and iron.

Sources

Amla 600mg/100g, guavas and cashew apples contain 200 to 300 mg /100g. Citrus fruits, such as oranges. Sweet lime, grape-fruit and pomelo are also excellent sources of ascorbic acid.

Drumsticks also have a high ascorbic acid content. Leafy vegetables, such as drumstick leaves, amaranth, cabbage, etc.,

Ascorbic acid occurs widely in plant foods, particularly in fresh fruits and vegetables become dry. Cut exposed to air most of the vitamin C originally present is destroyed. Fresh raw vegetables should be used for obtaining enough Vitamin C.

Sprouted pulses are good sources - 17-20 mg of vitamin C is produced during germination per 100 g of pulses.

We can get more vitamin C in your diet by:

- Harvesting at the peak of maturity.
- Storing in a cool, moist place
- Limiting exposure to air and sunlight
- Avoiding soaking food in water
- Cooking food in the minimal amount of water or even better, cooking by microwave.
- Cooking the food in pieces as large as possible.

Deficiency: Faulty cooking habits and inadequate intake of fresh vegetables and fruits are the major causes of dietary deficiency of vitamin C.

Scurvy is the most severe form of vitamin C. This is characterized by spongy bleeding gums, delayed wound healing and anaemia,

Recommended dietary allowance: Ascorbic acid cannot be stored in the body. The requirement has to be supplied daily, 40 mg per day for adult.

Large doses of vitamin C are prescribed during infections to increase the capacity of healing and immune competence. Vitamin C is a potent antioxidant and hence, large doses have been advocated for prevention of cancers and cardiovascular disease. It can lower blood concentrations of cholesterol and triglycerides.

7. Minerals

Major and Trace Minerals

Minerals are inorganic substances, unlike carbohydrates, fats, proteins and vitamins, which are organic compounds.

Minerals have two distinct characteristics: 1. Mineral elements do not provide energy 2. Mineral elements are not destroyed during cooking. Of this, 90 per cent is accounted for by seven minerals (calcium, phosphorus, potassium, sulphur, sodium, chloride and magnesium). The other minerals are known as trace elements. Although trace elements are present in very small quantities (iron, copper, selenium, zinc) in the bones and teeth. Blood and minerals are also found in soft tissues, such as nerves, muscles, blood and other body fluids.

General functions of minerals

1. Regulation of body functions:
   - Vitamin B_{12} contain cobalt
   - The enzyme carbonic anhydrase contains zinc
   - Hormone thyroxine contains iodine.
   - Incorporation of iron into haemoglobin needs copper
   - Synthesis of insulin in the pancreas needs zinc.
   - Muscle contraction and response of nerves are regulated by sodium, potassium, calcium and magnesium.

Absorption of mineral elements is favoured by Body requirement: If the body needs are high, as in pregnancy and growth stage of children, there is higher absorption of calcium and iron (up to 60 per cent) from the diet.

Stomach acidity: Absorption of calcium and iron is improved by the presence of acid conditions. Thus, absorption of calcium in the intestinal tract is enhanced by lactose, which is hydrolysed to lactic acid. The absorption of calcium, iron and zinc is improved by ascorbic acid.

Form of mineral is present in the food: Haem iron in animal foods is more easily absorbed in the body than iron compounds present in plant foods.

Absorption of mineral elements is reduced by

Some dietary components, which combine with mineral elements to form insoluble complexes, reduce their absorption. These components include oxialic acid, phytic acid, excess fiber.

The absorption of some minerals is adversely affected by an excessive intake of other minerals. For example, intake of iron supplements hinders the absorption of zinc, while zinc supplements reduce the absorption of copper.

Increased mobility of intestinal tract due to diarrhoea, intake of laxatives, etc., reduces the time of absorption and hence the amount absorbed.

Parasites present in the intestinal tract also interfere with the absorption of minerals. Iron deficiency anaemia in children is attributed to this factor.

Calcium

Functions

- Bones and teeth contain about 99 per cent of calcium in the body, in combination with phosphorus, protein and other minerals. These give the skeleton structure and rigidity.
- The remaining 1 per cent of calcium is present in the soft tissue and blood and is responsible for many regulatory functions. These functions, which it carries out along with other minerals are:
  - It is involved in normal muscle contraction, which includes heartbeat.
  - Control of the transmission of nerve impulses.
  - Maintenance of permeability of cell membranes to permit movement of material in and out of the cells.
  - Help normal clotting of blood.
  - Activate action of enzymes; e.g., pancreatic lipase as a factor. Ensure the absorption of vitamin B_{12}.

Food sources

The best source of calcium is milk. Milk products, such as curd, paneer, khoa, skim milk powder are good sources of calcium.

The millet ragi, sesame seeds (till) and green leafy vegetables are a good source of calcium.

Factors which favour the utilization of calcium are presence of vitamin D, ascorbic acid and lactose.

Efficient utilization during periods of rapid growth, pregnancy and childhood.

The presence of substances which form insoluble salts of calcium, e.g., Oxalic acid (in leafy vegetable). Phytic acid (in cereal bran) Excess of fat in the diet, may hinder calcium utilization. Excess of fiber in the digestive tract may increase faecal excretion of calcium, thus reducing utilization.

If adequate calcium is not provided during pregnancy and lactation, the bones of the women get porous (oesteoporosis).

In children lack of calcium leads to Rickets, which is a manifestation of vitamin D and/or calcium deficiency, is characterized by bowed legs, enlarged joints, beaded ribs and other deformities.

Table 13: RDA of Calcium (mg/day)

Group	Calcium
Adult men	600
Adult women	600
Pregnancy and lactation	1200
Infants	500
Children	
1-9 years	600
10-15 years	800
16-18 years	800

The best way to prevent osteoporosis-

- To maintain adequate calcium intake
- Physical activity to reduce loss of bone mass with age
- Exposure to sunlight to ensure vitamin D synthesis to aid calcium absorption.

Phosphorus

Bones and teeth contain 85 per cent of the phosphorus.

Functions of Phosphorus are

- Body building as an important component of bones and teeth.
- Fat absorption and transport
- As a part of ATP (adenosine triphosphate) and ADP (Adenosine diphosphate), which are essential for energy metabolism.

- As a component of enzymes needed in carbohydrate, fat and protein metabolism.
- As a part of buffer salts, which maintain acid-base balance in the body.

Food sources: The diet which meets the body's needs for calcium and proteins will also meet the body's needs of phosphorus. Rich sources of phosphorus include milk, eggs, flesh foods, legumes and nuts. Many processed foods and soft drinks have added phosphates.

Deficiency : Since phosphorous is so widely distributed, in the food stuffs its deficiency is quite rare.

Iron

Iron is a trace element. Amount of iron in the body is only about two to three grams.

Functions

The most important function of iron is transport and storage of oxygen:

- Iron combines with protein for the development of haemoglobin, the red pigment of the blood. The main function of iron in haem is to carry oxygen from the lungs to the cells and to carry back some of the carbon dioxide formed, to the lungs for exhalation.
- **Myoglobin** is found only in muscle, where it serves as a reservoir of oxygen.

Essential constituent of many tissues

- **Enzymes** which regulate redox function in the body.
- **Ferritin:** Iron containing protein in the liver, spleen and bone marrow.

Food sources

Iron from animal foods is better absorbed than from plant foods. Meat, fish, poultry, egg yolk. Liver and bone marrow very rich source (Milk is a poor source).Good sources of iron are dry fruit, whole grains, millets, pulses and green leafy vegetables. Foods processed in iron pans also contain appreciable quantities. Iron content of food can be increased by cooking in iron vessels.

Absorption

- The availability of dietary iron in the body is promoted if it is in a soluble form.
- Vitamin C favours iron absorption by dissolving the iron present.
- The presence of ascorbic acid helps iron absorption.
  - A high cereal diet, which has high phytate, low calcium and low ascorbic acid tends to make much of the iron present unavailable to the body.
  - Polyphenols: Present in coffee, tea, cocoa and some vegetables reduce the absorption of iron.
  - Severe infections - parasitic - hinder iron absorption.

Deficiency

Iron deficiency anaemia is widespread in India, 70 per cent in women and children. Etiological factor - is dietary iron deficiency, folate deficiency.

- The lack of iron results in anaemia due to insufficiency of haemoglobin.
- Hemochromatosis genetic defect, there is excessive absorption of iron resulting organ damage and skin pigmentation
- Hemosiderosis - Excess iron storage minus tissue damage. Abnormal breakdown of red blood corpuscles (RBCs) can also be the cause of haemosiderosis.

Table 14: Recommended dietary allowances of Iron (mg/ day)

Group	Iron mg/day
Man	17
Woman	21
Pregnancy	35
Lactation	25
Boys 16-18 years	28
Girls 16-18 years	26

Sodium and Potassium

Regulation of water balance : Sodium and potassium are responsible for the regulation of water balance between the inside and outside of cells. Potassium is mainly found in the fluid inside (intracellular fluid) the cell, while sodium is chiefly found in the fluid outside the cell (Extracellular fluid).

Maintenance of Acid - Base Balance : Important to maintain a constant pH in the body of 7.35 and 7.45. There is a continuous turnover of minerals in body tissues; they take up minerals, synthesise new compounds and discharge waste matter, which is known as dynamic equilibrium. This is regulated by hormones.

Iodine

Iodine is present in the thyroid gland. Two hormones produced by the thryroid gland, triiodothyroxine (T3) and thyroxine (T4) contain iodine. These hormones monitor the rate of energy metabolism in the body.

A deficiency of iodine is marked by a swelling of the neck enlargement of the thyroid gland. The condition is known as simple goiter.

Fluorine

Present in bones and teeth. The chief source is usually drinking water, which if it contains 1 part per million (ppm) of fluoride, supplies 1-2 mg/day.

Its presence in water > 2-3 ppm leads to fluorosis resulting in hardening of bone (Skeletal Fluorosis occurs as an endemic disease. The clinical manifestation of fluoride toxicity is called "genuvalgum" (knock knees). One important strategy to control fluorosis in these areas is to supply low fluoride water.

Where the fluoride content of the water is high (> 3-5 ppm) mottling of the teeth is common (Dental Fluorosis). While high intake of fluoride is a health hazard, a minimum level of fluoride in water (1-2 ppm) is essential to protect against dental caries.

To prevent fluorosis and dental caries, fluorine levels of water should be> 1 ppm but less than < 2 ppm.

Zinc

Zinc performs many functions as a part of every cell in the body and so, zinc is essential for normal growth, development, reproduction and immunity. Severe zinc deficiency is Acrodermatitis Enteropathica - a genetic disease - zinc absorption is significantly impaired.

ICMR has prescribed 15 mg/day as RDA for an Indian adult man. The body's need for zinc is greatest during periods of growth, infancy, adolescence, pregnancy and lactation 12 mg for men, 10 mg for women.

Sources

Soil zinc status determines the concentration of zinc in foods.

- Zinc from animal products is more readily absorbed than zinc from plant products.
- Meats, poultry, eggs and dairy product are the best sources.
- Outer layer of grains also contribute to zinc.
- Nuts contain high amount of zinc, fruit and most vegetables are fair sources.

8. Flavours

The enjoyment and acceptance of a meal depends to a large extent on the blend of different flavours present in food. Flavour is the most important attribute of food. It is detected by the senses of odour and taste. These stimulating components are synthesized in plants and animals. They may be further modified by cooking and processing. The intrinsic flavour of food represents the complex effect of these aromatic substances on our senses of odour and taste.

Definitions

Flavouring : It is a substance which can impart flavour and is generally used to impart taste or odour or both to a food.

Aroma : It refers to apleasant, often spicy odour, fragrance, or smell.

Flavour It is a blend of taste and odour perceptions experienced, when food is in the mouth.

Aftertaste : It is the flavour that lingers in the mouth after food has been swallowed.

Flavour intensifier It is a compound that enhances the flavour of other foods without contributing any flavour of its own.

The flavour of any food depends upon minute quantities of 100 or more chemicals that are present in food. These flavouring components are present in concentrations ranging from a few ppm to 0.1per cent. Food flavour are broadly classified as natural food flavours. Processed flavours, and added flavours which may be natural extracted flavours or synthetic flavours.

Natural flavours: These are usually extremely complex mixtures of many different substances. Sometimes, the flavour of a natural flavouring agent may depend upon a single substance, e.g., the flavour of clove oil is because of the chemical eugenol that constitutes 85 per cent of clove oil or it may be present in extremely small amount, such as citral in oil of lemon.

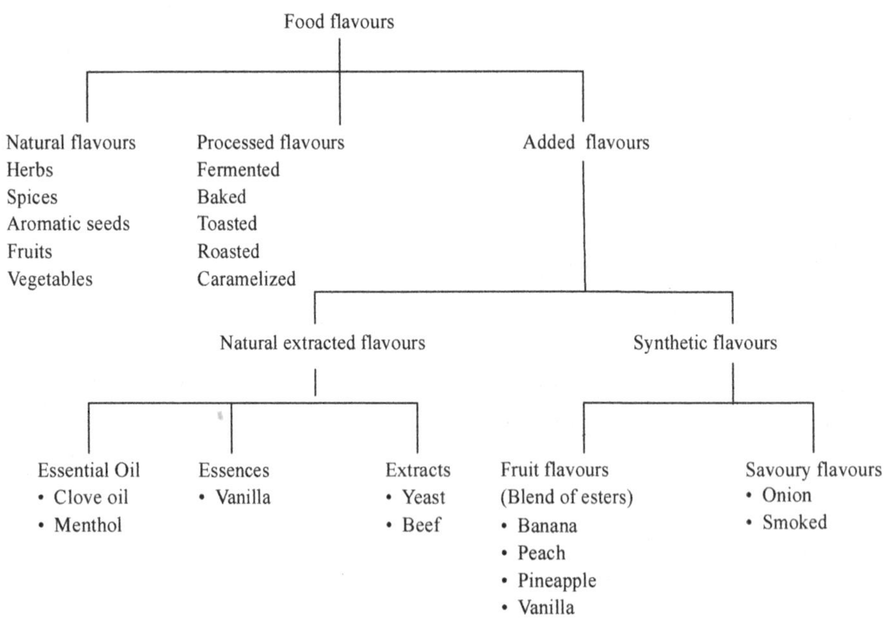

Fig. 7: Classification of flavours

9. Food colours

Colours are added to many food items, to improve their appearance and to give the public appetizing and attractive qualities they desire. Colours from natural materials, such as annatto, caramel, carotene and saffron are the best examples. Colours of synthetic origin which include coal tar dyes. Synthetic origin, which include coal tar dyes, have been examined for their safety. Synthetic colours generally excel in colouring power, uniformity, stability and lower cost. Carbonated beverages, candies, gelatin desserts and bakery goods are among items coloured with certified coal-tar dyes. Food colours also include inorganic materials, such as iron oxide to give redness and titanium dioxide to intensity whiteness. In India, natural pigments, their extracts of synthetic equivalents, *viz.* beta-carotene, bet-apo-8 carotenic acid, canthaxathin, (turmeric) are permitted to be added in food. Among the synthetic dyes, permitted colours are given in Table 15.

Stabilizers and thickeners

These include gums, gelatin, starches, dextrins, protein derivatives and other additives that stabilize and thicken food by combining with water to add viscosity and to form gels. Gravies, pie fillings, cake toppings, chocolate milk drinks, jellies, puddings and salad dressings are among the many foods.

Classification

Table 15: Permitted food colours

Natural or synthetic equivalent	Artificial synthetic dyes
Beta carotene	Panceau-42
Beta-apo-8′-carotene	Carmosine
Esters of Beta-apo-8′-carotenic acid	Fast red E
	Amaranth
Canthaxanthin	Erythrosine
Chlorophyll	Tartrazine
Annatto	Sunset yellow
Caramel	Indigo carmine
Ratanjot	Brilliant blue
Saffron	Green
Curcumin (turmeric)	Fast green

10. Bioactive compounds

A bioactive compound is a compound that has an effect on a living organism, tissue or cell. In the field of nutrition, Bioactive compounds are distinguished from essential nutrients. While nutrients are essential to the sustainability of a body, the bioactive compounds are not essential since the body can function properly without them, or because nutrients fulfill the same function. Bioactive compounds can have an influence on health.

Bioactive compounds are also defined as a type of chemical found in small amounts in plants and certain foods (such as fruits, vegetables, nuts, oils, and whole grains). Bioactive compounds have actions in the body that may promote good health. They are being studied in the prevention of cancer, heart disease and other diseases. Examples of bioactive compounds include lycopene, resveratrol, lignan, tannins, and indoles.

Origin and examples: Bioactive compounds are found in both plant and animal products or can be synthetically produced. Examples of plant bioactive compounds are carotenoids and polyphenols (from fruits and vegetables), or phytosterols (from oils).

Example in animal products are fatty acids, found in milk and fish.

Some examples of bioactive compounds are flavonoids:

- Caffeine
- Carotenoids
- Carnitine
- Choline
- Coenzyme Q
- Creatine
- Dithiolthiones
- Phytosterols
- Polysaccharides
- Phytoestrogens
- Glucosinolates
- Polyphenols
- Anthocyanins
- Prebiotics
- Taurine

Health

Sellers of bioactive substances often attribute health benefits to these compounds, but there is insufficient research into the effectiveness and safety of these substances, either in long-term use or in quantities that exceed normal consumption levels. In addition, some flavonoids have been shown to influence the effects of drugs. However, a number of bioactive substances have been shown to act as an antioxidant· As bioactive compounds are not essential, advice on daily intake is often unregulated.

Food sources and Bioactive compounds

1. **Allium sulphur compounds:** Leeks, onions, garlic
2. **Anthocyanins:** Brinjal, grapes, berries
3. **Beta carotene:** Pumpkin, mangoes, apricots, carrots, spinach, parsley
4. **Catechins:** Red wine, tea
5. **Copper:** Seafood, lean meat, milk, nuts, legumes
6. **Cryptoxanthins:** Red peppers, pumpkin, mangoes
7. **Flavonoids:** Tea, green tea, red wine, citrus fruits, onion, apples
8. **Indoles:** Cruciferous vegetables, such as broccoli, cabbage, cauliflower
9. **Lignans:** Sesame seeds, bran, whole grains, vegetables
10. **Lutein:** Corn, leafy greens (such as spinach)
11. **Lycopene:** Tomatoes, grapefruit, Watermelon
12. **Manganese:** Seafood, lean meat, milk, nuts
13. **Polyphenols:** Thyme, oregano
14. **Selenium:** Seafood, offal, lean meat, whole grains
15. **Vitamin C:** Oranges, berries, kiwi fruit, mangoes, broccoli, spinach, peppers
16. **Vitamin E:** Vegetable oils, nuts, avocados, seeds, whole grains
17. **Zinc:** Seafood, lean meat, milk, nuts
18. **Zoochemicals:** Red meat, offal, fish.

❑❑❑

Chapter 3

Food Microbiology

Introduction to Food Microbiology

Food microbiology is a branch of microbiology concerned with the relationship between micro-organisms and food. It covers food borne diseases, food hygiene, food spoilage, fermented foods and beverages, use of micro-organisms to produce food ingredients, microbial aspects of quality control, microbiological analysis of food and food legislation.

Bacteria : They are single celled organisms with typical dimensions of around 1 micro metre. Bacteria are the most important and well studied food borne pathogens. A key factor is their ability to rapidly multiply in food thus increasing the risk of food borne illness. Most of them are grown in their temperature range from 20 to 45°C (68 to 113^0 F). However, certain food borne pathogens (termed psychrotrophs) are capable of growth under refrigerated conditions or temperatures less than 10^0 C (50^{0}F). Bacteria also vary in their resistance to high salt, sugar or total solids level as well as the acidity of food products.

Molds: Mold growth on foods, with its fuzzy or cottony appearance, sometimes colored, is familiar to everyone and usually food with a moldy or "mildewed" food is considered unfit to eat. Special molds are useful in the manufacture of certain foods or ingredients of foods. Thus, some kinds of cheese are mold-ripened, e.g., blue, Briemolds are used in making oriental foods, e.g., Soya sauce, miso. Molds have been grown as food or feed and are employed to produce products used in foods, such as amylase for bread making or citric acid used in soft drinks. Some molds do produce various toxic metabolites (mycotoxins).

Yeasts and yeast like fungi: Like mould, the term "yeast" is commonly used but hard to define. It refers to those fungi which are generally non-filamentous but unicellular and ovoid or spheroid and which reproduce by budding or fission.

Yeasts may be useful or harmful in foods. Yeast fermentations are involved in the manufacture of foods, such as bread, beer, wines, vinegar and surface ripened cheese and yeasts are grown for enzymes and for food. Yeasts are undesirable when they cause spoilage of sauerkraut, fruit juices, syrups; molasses, honey, jellies, meats, wine, beer, and other foods.

Causes of Spoilage of food by micro-organisms

Spoilage may be due to one or more of the following factors:

1. Growth and activity of micro-organisms
2. Insects
3. Action of the enzymes of the plant or animal food.
4. Purely chemical reactions i.e., those not catalysed by enzymes of the tissues or of micro-organisms
5. Physical changes, such as those caused by freezing, burning, drying, pressure, etc.

Classification of foods by ease of spoilage

1. Stable or non-perishable foods

These foods which do not spoil unless handled carelessly include such products as sugar, flour and dry beans.

2. Semi-perishable foods

If these foods are properly handled and stored, they will remain unspoiled for a fairly long period. e.g. Potatoes, apples, etc.

3. Perishable foods

It includes most important daily foods that spoil readily e.g. Meats, fish, poultry, milk, eggs, fruits and vegetables.

Factors affecting growth and survival of micro-organisms in foods

1. Intrinsic factors

Factors related to the food itself like:

- **Nutrients -** In order to grow and function normally the micro-organisms of importance in foods require the nutrients like water, source of energy, source of nitrogen, vitamins and related growth factors and minerals.
- **Water activity -** It is the index of the availability of water for microbial growth.
- **pH** : It is the negative logarithm of the hydrogen ion activity. Every micro-organism has a minimal, a maximal and an optimal pH for growth. Bacteria grow fastest in the pH range 6.0-8.0, yeasts 4.5-6.0 and filamentous fungi 3.5-4.0. Usually between pH 5.0 & 6.0.
- **Redox potential (Eh)** - Micro-organisms display varying degrees of sensitivity to the oxidation-reduction potential of their growth medium. Eh of substance is defined as the ease with which the substrate loses or gains electrons. When an element or compound loses electrons, the substrate is said to be oxidized, where as a substrate that gains electrons becomes reduce.

2. Extrinsic factors

Factors in the environment and external to the food, which affect both the micro-organisms and the food itself during processing and storage.

Extrinsic factors include

- **Storage temperature-Foods are exposed to different temperatures from time of production until the time of consumption.**
    - Microbial growth is accomplished through enzymatic reactions which is depended on temperature.
    - Microbial Groups: Psychrophiles, Mesophiles & Thermophiles
    - Every 10°C rise doubles the Catalytic rate of enzyme and every 10°C decrease reduces it to half.

- **Relative humidity- Relative humidity is a measure of water activity of the gas phase.**
    - High RH influence the water content of food and high water content promote microbial growth.
    - When food stored in high relative humidity, dormant spores of bacteria or fungi to germinate.
    - Once they are actively growing, they produce water as an end product of respiration.
    - Hence, they increase the water activity of their own, this favours the growth of high water activity requiring bacteria and increase in spoilage of food.
    - Therefore, dry conditions are considered better for food storage as than moist conditions.
- **Gases - Presence or absence of gases affects microbial affects type of microbial populations**
    - Carbon dioxide regulates cell growth of certain bacteria.
    - If partial pressure of carbon dioxide increases over a critical level, metabolic activity will be retarded.
    - Delaying effect of Carbon dioxide increases with increase in concentrations.
    - Carbon dioxide is used in packaging of some food items in order to control the growth of microbes.
- **Activities of other micro-organisms - Presence or absence of other microbes affects type of microbial populations.**
    - Microbial competition in food.
    - Some microbes produces antimicrobial substances, such as antibiotics, organic acids, alcohol and bacteriocins.
    - Those metabolites inhibit the growth of other microbes.

3. Production of fermented foods

Fermentation is one of the oldest forms of food preservation technologies in the world. Indigenous fermented foods, such as bread and wine have been prepared and consumed for the thousands of years and are strongly linked to culture and tradition, especially in the rural households and village communities. The development of fermentation technologies is lost in the midst of history.

Fermentation is the slow decomposition process of organic substance induced by micro-organisms or by complex nitrogenous substance (enzymes) of plant or animal origin. It can be described as a biochemical change, which is brought about by the anaerobic or practically anaerobic oxidation of carbohydrates by either micro-organisms or enzymes. Micro-organisms are naturally found in foods, since there is no environment where some type of microbes cannot live. These microbes either living or dead and their cellular byproducts all have specific uses in some foods. These include such products as fermented food products. Fermented foods use microbes to convert the original food into a fermented product by the use of specific microbes. These micro-organisms use the original product for growth and reproduction and in the process they excrete byproducts into the environment surrounding themselves and food. These byproducts plus the part of the original product that is not consumed is fermented food. Fermented foods include fermented dairy, meat, fish, cereals, fruits, vegetable products, etc. they made be fermented separately or in conjunction with each other to produce the desirable and product.

Fermentation involves the introduction of the desirable microbes into the original products. Some of the common microbes used in food fermentation are highlighted herewith:

- *Lactococcus lactis* - used in dairy fermentation
- *Streptococcus thermophiles* - used in dairy fermentation
- *Leuconostoc* sp. - used in wine making, dairy fermentation
- *Pediococcus* sp. - meat fermentation, vegetable fermentation, ripening of some cheeses.
- *Lactobacillus* sp. - meat fermentation, vegetable fermentation dairy fermentation, sourdough bread.
- *Bifidobacterium* sp. - added to dairy products to promote intestinal health
- *Propionibacterium* sp. - Swiss cheese.
- *Yeasts* - bread, beer, wine, liquors

- *Moulds*– ripening cheese, soya sauce.
- *Lactobacillus delbruekii, species : bulgaricus and streptococcus thermophiles* – making of yogurt

Although fermentation of foods has been in use for thousands of years, it is likely that the microbial and enzymatic process responsible for the transformations were largely unknown. It is only recently that there has been a development in the understanding of these processes and their adaption for commercialization.

i) Idli

Idli is prepared from rice and black gram dal. They are cleaned, washed and ground in equal proportions and left overnight for fermentation. The micro-organisms involved in this preparations are *Leuconostoc mesenteroides,* which grows first in the batter followed by *streptococcus faecalis* and *Pediococcus cerevisiae*. Once the batter rises sufficiently, it is steam-cooked and served.

ii) Dosa

Dosa is a light, shallow fried, thin pan cake. It is prepared from fermenting rice and black gram overnight. The aeration of batter is caused by lactic acid bacterial fermentation by *Streptococcus faecalis* with carbon dioxide production *by Leuconostoc mesenteroides.*

iii) Dhokla

Dhokla is similar to idli, where rice and bengal gram are used in making it yellow coloured. *Leuconostoc mesenteroids* and *Streptococcus faecalis* are the bacteria involved in the fermentation. The batter is poured into large sheets, then steamed and cut into pieces and seasoned. Apart from preparing this foods, the micro-organisms have contributed to a great extent in preparing man's desire for alcoholic beverages, which are the products of plant fermentation.

Economically important fermentation products

Alcoholic drinks fall into two broad categories: wines and beers. Wines are made from the juice of fruits and beers from cereal grains.

i) Beer

Beer is an alcoholic product, produced by brewing. It is a principal malt beverage where the fermentation of carbohydrates to alcohol takes place. Barley is used in the preparation of beer. Yeasts play a major role in the preparation of beer. Large beer is produced by *Saccharomyces uvarum,* which settles at the bottom of fermenting vat and is known as the 'bottom yeast'.

ii) Wine

Wine is one of the oldest and well known fermented alcoholic beverages produced by the fermentation of good and sound grapes which is further processed, known as aging, before consumption. Although there are other fruit wines, they are not as popular as grape wine. The grape fruits are crushed to give a "must". The fermentation of "must" is initiated by yeast *Kloeckera apiculatus* and *Methschnikowia Pulcherrima* together with the yaest of *Torulopsis,* Candida, etc. the main fermentation is by the yeast, *Saccharomyces cerevisiae* and S *uvaum* or S. *bayanus* which converts sugars to ethanol and carbon dioxide.

The different microbes used in the food industry to manufacture the products of commercial value, namely vinegar citric acid antibiotics, etc.

Vinegar

If the alcohol produced by the fermentation process is further oxidized to acetic acid by the acetic acid producing bacteria, the product is vinegar. Vinegar is made by different process. It can be made from fruit juices, starchy vegetables, malted cereals, sugars and alcohol. Vinegar used as preservative in food preparation.

□□□

Chapter **4**

Principles and Methods of Food Processing and Preservation & Its Advantages

Food Processing

Food processing is the transformation of cooked ingredients, by physical or chemical means into food, or of food into other forms. Food processing combines raw food ingredients to produce marketable food products that can be easily prepared and served by the consumer. Food processing typically involves activities, such as mincing and macerating, liquefaction, emulsification and cooking (such as boiling, broiling, frying or grilling); pickling, pasteurization and many other kinds of preservation; and canning or other packaging. (Primary-processing, such as dicing, slicing, freezing or drying, when leading to secondary products are also included).

Food Processing includes conversion of a produce into a product with added value by different basic processing of dehydration, pulp extraction, powder making, cooking, frying, etc. To get a condition that is more finished, stored or marketable. e.g. Cereal flour, Turdal, Fruit Jam, etc.

Food Preservation

Food preservation prevents the growth of micro-organisms (such as yeasts), or other micro-organisms (although some methods work by introducing benign bacteria or fungi to the food), as well as slowing the oxidation of fats that cause rancidity. Food preservation may also include processes that inhibit visual deterioration, such as the enzymatic browning reaction in apples after they are cut during food preparation.

Many processes designed to preserve food involve more than one food preservation method. Preserving fruit by turning it into jam, for example, involves boiling (to reduce the fruit's moisture content and to kill bacteria, etc.), sugaring (to prevent their re-growth) and sealing within an airtight jar (to prevent recontamination). Some traditional methods of preserving food have been shown to have a lower energy input and carbon footprint, when compared to modern methods. Food preservation can be defined as the science, which deals with the process of preservation of decay or spoilage of food and helps it to be stored in a fit condition for future use.

Food Spoilage

Food can spoil at any stage in its preparation or storage. In fact, spoilage in food begins as soon as :

1. Vegetables and fruits are harvested
2. Eggs are laid
3. Fish is caught
4. Animals are slaughtered for meat
5. Milk is drawn from milch animals

This spoilage continues till food is consumed.

Spoilage : Spoilage can be defined as decomposition or damage causing undesirable changes in food. It is caused due to various agents making food unsuitable for consumption. A spoiled food looks, smells, and tastes bad.

Contamination: The term 'contaminated' is used for those foods which are not fit to be eaten for sanitary reasons. Although contaminated food may look, smell, and taste good, it may contain harmful chemicals, non-food matter and bacteria. Contamination of food results in its spoilage.

Causes of food spoilage

Food spoils because of any one or more of the following reasons:

Growth and activity of bacteria, yeast and mould: Micro-organisms can cause visible changes in the food. Milk turning sour, mould growth on bread and fruit juices fermented by yeast are some examples of visible signs of spoilage. The kind of micro-organisms which spoil the food will depend on the composition of food, i.e., its pH, moisture content, nutrients, temperature, etc.

Insect infection: Insects, such as worms, weevils and moths infest cereal grains and make grains unfit for consumption. Weevil bore holes into grains and multiply, destroying the flavour and the grain.

Enzymatic changes: Natural enzymatic changes by autolysis, such as over maturing, softening, browning and sprouting, damage the food. Sprouting of potatoes, browning/darkening of bananas and softening of fruits and vegetables are seen after harvesting, if cool temperatures are not maintained.

Chemical action: Chemical reactions which are not catalysed by enzymes or micro-organisms can also result in chemical spoilage of food. Oxidative rancidity in fats and hydrogen swell in canned foods are examples of this type of spoilage.

Physical changes: Physical changes in food caused by freezing, dessication, absorption of moisture, etc., can spoil food. Mechanical damage, such as bruising or cracked egg shells can accelerate spoilage by micro-organisms or by enzymes. Bruised apples which are brown when cut, freezer burn in deep frozen foods or cracked eggs are signs of spoilage in foods.

Spoiled food cannot be rectified by any processing method. It results in wastage and should be discarded immediately. Only food which is wholesome, i.e., at the desired stage of maturity, free from pollution, and any objectionable changes resulting from microbial or enzyme action should be preserved.

Markets are flooded with a vast array of processed foods, which are processed using latest technology, to meet the year round demand during the lean period.

Importance of food preservation

Preservation of food material has played a vital role avoiding widespread starvation and it continues to do so till date, even in industrially developed world. Food supply has to keep pace with the needs of the population. There is always shortage of food in developing countries like India because of demand of increasing population. Therefore, it is important to improve and expand facilities for storage and preservation of food.

Delay in the use of fresh foods alters their freshness, palatability and nutritive value but by preserving these foods, spoilage can be avoided. Thus, perishable foods can be made available throughout the year by preservation.

By preservation and proper storage of food, it can be saved for future use at the time of scarcity, natural drought, etc.

Preservation of food also helps in stabilizing the prices of the food by making the availability of seasonal foods throughout the year.

Objectives of food processing

The main objectives of food processing on a home scale, or in the food industry are similaras indicated below:

1. **Removal of unwanted matter from the food:** Unwanted matter may be inedible, indigestible or harmful to health, such as husk from grains, rind from oranges, cherry stones and coconut shells. Unwanted matter is removed by appropriately designed gadgets. The processes include shelling, destoning, milling, peeling, *etc.*

2. **Making food safe for consumption:** Some foods contain natural toxins which need to be inactivated, *e.g.*, trypsin inhibitors in soyabeans Fungal toxins, such as aflatoxin in moldy groundnut and grains, infected portions of food material and green portion in potatoes are removed by visual examination, and chemical toxins and poisons are discarded. Ensure safety of food by using processes to remove toxins, and heat to destroy microorganisms and their toxins. Develop safe processes to prevent contamination at any stage.

3. **Increasing digestibility**: Most foods are difficult to digest unless they are cooked. Cooking softens fiber, gelatinizes starch, dentures protein and makes food easier to digest. With the exception of fruit, all other foods need some kind of processing to make them more digestible.

4. **Enhance flavour, colour and taste**: The acceptability of food depends to a large extent on its organoleptic or sensory qualities. Processing techniques enhance the appearance of food and many techniques make food more appetizing and tasteful, such as baking of cakes and bread. The brown crust is formed due to Maillard reaction which gives bakery items its characteristic baked flavour, aroma and taste. Processes, such as

Caramelization of sugar, fermentation of idli batter and alcoholic fermentations produce a superior taste as compared to unprocessed foods.

5. **Improving texture and consistency:** Processes, such as emulsification, aeration, gel formation and increase in viscosity are aimed at improving the texture and consistency of ready-to-eat, cook-chill and cook-freeze operations. Processing prevents changes in consistency of such operations during the freeze-thaw process. Special starches which do not retrograde or show syneresis are used for setting gels. Crystal formation in fondants, jams, and ice cream can be retarded in processed foods and desirable texture can be obtained.
6. **Minimizing nutrient loss**: Nutrients are better retained by controlled processing conditions, such as autoclaving, freeze drying and controlled heat. Nutrients are better retained by controlled processing. Processed margarine a substitute for butter, is fortified with 30 IU vitamin A and 2 IU vitamin D per gram of fat. Processed foods are often enriched with vitamins, minerals and lysine.
7. **Extending the shelf life** : Processing extends the shelf life because apart from removing unwanted, spoilt and harmful matter and subjecting the food to temperatures outside the danger zone, all processes, such as dehydration, cold storage, canning and pasteurization are aimed at preservation of food.
8. **Increasing acceptability through fabricated foods.** New products of uniform size and shape are being introduced in the market. They are made from low-grade commodities which are plentiful or good for health.

Principles of food preservation

Food spoilage may be caused due to chemical and biological factors. Inherent enzymes present in food may cause spoilage under certain conditions. Growth of bacteria, molds or yeasts already present or inoculated in food while processing causes more serious spoilage than enzymes. To prevent spoilage of food, various methods of preservation are available which are based on certain principles.

The basic principles of food preservation primarily involves the process of inhibition of micro-organisms:

- The growth and activity of micro-organism.
- Activity of endogenous enzymes.
- Invasion and spoilage by insects and rodents.
- Inhibiting the growth and activity of microorganism: This can be achieved by keeping out micro-organism (Asepsis)
- Removing the micro-organism, *e.g.* by filtration
- Hindering the growth and activity of micro-organisms, *e.g.* by the low temperature, drying, anaerobic conditions, chemical or antibiotics.
- Killing the micro-organism. *e.g.* by heat, or radiation.
- Inhibit the activity of endogenous enzymes: This can be achieved by: Destruction or inactivation of food enzymes, *e.g.* by blanching.
- Prevention or delay of purely chemical reaction, *e.g.* prevention of oxidation by use of antioxidants.

Classification for food preservation

Foods have been preserved for centuries by various techniques. Various methods for the preservation of foods based on techniques which eliminate moisture and other factors causing food spoilage have been devised and practiced. For purpose of food preservation, foods are classified in perishable, semi-perishable and non-perishable:

- Perishable foods
- Semi-perishable foods
- Non-perishable foods

Methods of Food Preservation

1. Use of high Temperatures

High temperatures destroy micro-organisms by denaturation of cell proteins and inactivation of enzymes needed by them for their metabolism. At temperature above 63°C, bacteria stop multiplying and as the temperature increases, they are gradually destroyed. The thermal death time (TDT) is the time needed at a given temperature to kill a number of microbes. Heat used to destroy microbes may be in the form of wet heat or dry heat.

a) Pasteurization

Pasteurization is a heat treatment that kills a part but not all the micro-organisms present and usually involves the application of temperatures below 100°C. This process serves two objectives:

1. It destroys pathogenic organisms associated with the food.
2. Extends the products shelf-life by decreasing the microbial population and inactivating some enzymes.

There are three methods of pasteurization:

1. Bottle or "Holding" pasteurization

Bottle or "Holding" pasteurization: This method is commonly used for the preservation of fruit juices at home. The extracted juice is strained and filled in bottles. The bottles are then sealed air-tight and pasteurized.

2. Overflow method

Overflow method : Juice is heated to a temperature of about 2.5°C higher than the pasteurization temperature, and filled in hot sterilized bottles. The sealed bottles are sterilized at a temperature of 2.5°C and then cooled.

3. Flash pasteurization

The juice is heated rapidly to a temperature of about 5.5°C higher than the pasteurization temperature and kept at this temperature for about a minute. This method has been developed specially for canning of natural orange juice, grape and apple juice. The advantages are loss of flavour is minimum, and vitamins are not destroyed.

The choice of temperature and time of pasteurization will be influenced by the consideration of the purpose of the process and the chemical and physical composition of food. Thus, milk is pasteurized at 62.8°C for 30 minutes, to inactivate. Pathogens and whole egg at 64.4°C for 2-5 minutes, to control the dissemination of *salmonella* sp. The other foods which are usually pasteurized are beer, wines, fruit juices, etc.

Pasteurized products are not sterile. They contain vegetative organisms and spores still capable of growth. Pasteurized foods must be stored under refrigeration.

The spore-forming bacteria found in tomato juice, however, require heating at a higher temperature for a much longer duration. Acid fruit juices require a lower temperature and less time for pasteurization than the less acid one. Juices and milk can be pasteurized in two ways:

HTST (high temperature-short time) method: HTST, using a temperature of 130°C and above for short times, for example, milk is heated to 72°C and kept at this temperature for 15 seconds.

LTLT (low temperature-long time) method: In this method, the product is heated at lower temperature for a longer time. For example, milk is heated at 62.8°C for 30 minutes.

b) Sterilization

The complete destruction of micro-organisms is known as sterilization of food also adversely affects the properties of foods, the heat treatment given should be just enough to destroy the pathogenic organisms and toxins and ensure desire storage life.

c) Blanching

Foods which are to be frozen, dried or canned are immersed in hot boiling water for a few minutes prior to processing. Blanching helps in removal of peel, inactivation of enzymes that oxidize vitamin D, removal of gas in tissue spaces and wilting of the tissue which helps in proper filling of the can. The enzymes which bring about discolouration or browning seen in apples, pears and potatoes are also inactivated.

d) Canning

Canning is the process in which the foods are heated in hermetically sealed. air tight) jars or cans to a temperature that destroys micro-organism and inactivates enzymes that would be a health hazard or cause the food to spoil.

The temperature used are above 100°C. It is used to preserve fruits, vegetables, fish, meat, etc., and no preservatives are added to the food.

Aseptic Canning

Aseptic canning, heat transfers from the outside of the container to the inside will require several minutes or even hours depending upon the container size to reach the sterilization temperature. The time of sterilization can be shortened to seconds or even fraction of a second in aseptic canning. The basic principle of this method is that food is pumped continuously through a plate-type or tubular heat exchanger which heats it very quickly to a high temperature holding it at that temperature for time required and then cooling. Food temperature employed may be as high as 150°C and sterilization takes place in 1 or 2 seconds. Such a rapid sterilization at high temperature is referred to as ultra-high-temperature (UHT) sterilization. The sterile food is quickly cooled, placed in aseptic containers and the lids are sealed which in a sterile environment. The food canned aseptically retains the nutrients and the sensory attributes will be good, Such types of foods if stored for one year may lose only 25% ascorbic acid from fruits and vegetables, 20-30% of thiamine from meat but riboflavin content is not affected. Carotene content losses in fruits and vegetables are small even after months of storage.

2. Use of low Temperatures

Low temperatures preserve food by retarding chemical reactions, enzymatic action, and growth and activity of micro-organisms. Lower the temperature, the better the food will be preserved. At low temperatures, micro-organisms are not killed and since foods already have a microbial load, micro-organisms will multiply once the temperatures become favourable.

a) Refrigeration / Chilling

Temperatures of 1°C to 4°C prevent food from spoiling for short periods. It merely retards the decay but frozen foods kept at - 18°C preserve food for a year and at - 28°C food can be preserved for upto two years.

Chilling temperatures retard microbial growth and biochemical changes which affect colour, texture, flavour, and nutritive value. Most perishable foods, such as eggs, dairy products, chicken, meat, and seafood are held at chilling temperatures for a limited time without affecting their condition. They need to be consumed by the 'best before date'. The relative humidity (RH) in refrigerator storage is another important factor which needs to be controlled. A low RH results in loss of moisture and wilting and shrinkage in fruits and vegetables. A high RH favours surface spoilage by micro-organisms.

Cook chill

In this system, food is cooked in the kitchen in advance and rapidly chilled and stored at 0-3°C. It is reheated just before it is served. The cook-chill system is used for almost all kinds of foods and is popular in airlines and institutional catering. Proper cooking procedures and hygienic standards need to be followed to ensure safety and quality.

a) Freezing

Food is preserved for long periods by reducing its temperature to - 18°C or lower. At this temperature, water present in food is converted to ice and microbial growth stops. Freezing retains colour, flavour and nutritive value. However, the texture of some foods is adversely affected. If properly stored, frozen food has a shelf life of 3-12 months, fruits, vegetables, meat, fish, and poultry can be preserved in this way.

Food to be frozen should be frozen quickly so that small ice crystals form in the cells of food which is desirable. Slow freezing causes large irregular shaped ice crystals to form in the cells. These contain the flavour and nutrients. When food is thawed, flavour and some nutrients are lost as large ice crystals pierce the cells. Food is quickly frozen by using any one of the following equipment:

Blast freezers : In this cold-air at - 18°C to 34°C is vigorously circulated over food while it passes through an insulated tunnel.

Plate freezers: Food to be frozen is placed in contact with a metal surface which is cooled by refrigerant. It is used for ice creams, juices, etc., both packaged and unpackaged foods.

Immersion freezers: Packaged or unpackaged food is frozen by immersing it or spraying it with a freezing agent. It is used for freezing poultry.

Spray Freezers: This is the quickest freezing method in which liquid nitrogen or carbon dioxide is used. It is called cryogenic freezing. The food to be frozen is placed on a conveyor belt which passes through an insulated freezing tunnel. Liquid nitrogen or carbon dioxide is injected into the tunnel through a spray which changes its state and vaporizes, resulting in instant freezing.

Freeze flow: In this system, food freezes but does not harden.

Cook-Freeze

It is a specialized method of a food processing system in which food is cooked and immediately blast frozen at - 18°C or below, and stored at that temperature till it has to be served. Such food has a shelf-life of 3 to 6 months. The food should be frozen immediately so that large ice crystals do not form and freshness is retained and should be accomplished within one-to-one and a half hours.

3. Use of Chemicals (Preservatives)

A preservative is any substance which retards deterioration of food. There are two groups of preservatives used in food:

a) Class I preservatives

This class includes sugar, common salt, glucose, alcohol, spices, vinegar or acetic acid, honey, and wood smoke. There is no restriction by law on the addition of these substances in food.

b) Class II preservatives

This class includes chemicals which inhibit microbial growth and can be added to certain foods only in definite permitted limits. They are usually added at the end of the processing operation and their presence has to be mentioned on the label, *i.e*, there is a restriction on the use of these preservatives in food by law. They include salts of benzoic acid, fumaric acid, sulphurous acid, nitrates and nitrites of sodium and potassium, sorbic, propionic, and acetic acid. Nisin, an antibiotic, is used in the preservation of cheese (**Table 16**).

They are used to preserve food in the following ways:

1. They may be added to food, e.g., sodium benzoate in tomato sauce and potassium metabisulphate in lemon squash.
2. They may be applied on the surface of foods. Sulphur dioxide is used on dry fruits and borax is used to wash vegetables and whole fruit.

3. Wrappers may be impregnated with sorbic acid to prevent surface spoilage of cheese.
4. The ice used to chill foods like fish may contain tetracycline, an antibiotic which is a permitted preservative.
5. They may be used as gases around food. Fruits and vegetables are stored in an atmosphere containing two to three percent carbon dioxide to retard the ripening process.

Table 16: Uses of chemical preservatives

Class I	Class II
They include sugar, salt, vinegar, spices, smoke, and oil.	**They include salts of various acids.**
a) Sugar: A high concentration of sugar prevents molds, yeast and bacteria and is used to preserve jams, jellies, marmalades, fruit preserves, candied fruit, glazed fruit, crystallized fruit, and chutneys.	a) Benzoic acid and its salts, e.g, sodium benzoate in tomato sauce, squashes, and syrups.
b) Salt: Micro-organisms cannot grow in high concentrations of salt. Meat is pickled in salt solution, e.g., ham and ox tongue. Salt is used for pickling and curing. Mango, lime, and vegetable pickles are preserved by salt. Salt and sugar preserve food by binding water and making it unavailable for microbial activity.	b) Sulphurous acid and its salts, *e.g.*, potassium metabisulphate in fruit beverages, such as lime cordial, before dehydration of fruits and vegetable, and in bulk storage of fruit pulp. c) Nitrates and Nitrites of potassium and sodium are used in the preservation of meat and for colour retention in pork.
c) Smoke: It is a complex mixture of alcohols, acids, phenols, and toxic substances which inhibit microbial growth. Meat to be smoked is first salted and then smoked. Temperature control during smoking is essential to maintain the texture.	d) Propionic acid and its salts and such as sodium and calcium propionate are used as mold inhibitors on cheese surface and to prevent rope in bread.

d) Vinegar: 4 per cent acetic acid in water is used for pickling vegetables, such as onions, gherkins, red cabbage, shallots, and to preserve chutneys.

e) Sorbic acid and its salts, such as potassium sorbate is used in fruit juice, sauces, jam, flour confectionery, and in magarine to prevent yeast & mold growth.

Controlled Atmosphere Storage

When fruits and vegetables are harvested, they are still alive and respire using up oxygen and giving out carbon dioxide, water vapour, and heat. The faster the rate of respiration, the quicker the ripening process. The ripening rate can be reduced by :

1. Lowering the temperature
2. Reducing available oxygen
3. Increasing the carbon dioxide concentration.

All three methods are used in controlled atmosphere storage.

Fruits can stored for varying lengths of time under controlled conditions, thus extending their shelf-life.

Modified atmosphere storage is also used for food storage. It differs from controlled atmosphere storage because in controlled atmosphere storage the percentage of gas is monitored. While in modified atmosphere the air is initially replaced with gas, but no further measures are taken to check the percentage of gas in the atmosphere. The concentration proportion of gases used varies with the kind and variety of fruit. The carbon dioxide concentration is usually increased and oxygen concentration is decreased for fresh fruits and vegetables stored in cold rooms.

4. Drying methods and its importance

a) Solar drying

i) **Sun drying:** Drying in the sun is very economical. You only have to spread the produce on a suitable surface and let it dry in the sun.

Transfer of the moisture from the surface to the surrounding air, in the form of water vapor. Solar radiation, in the form of solar thermal energy, is an alternative source of energy for drying especially to dry fruits, vegetables, agricultural grains and other kinds of material, as wood, etc.

Disadvantages

Somebody has to stay at home throughout the drying period to chase off domestic animals, to remove the produce, when the weather becomes too windy and dusty, or when it rains.

The dried product is often of poor quality as a result of grit and dirt.

The product is often unhygienic as a result of micro-organisms and insects, such as flies.

ii) Solar drying

The technology and capital required to dry fruit and vegetables by solar dryers is basic and the entire operation can be completed in most kitchens. The structure can be very basic, e.g. a box frame covered with plastic sheeting.

Advantages of solar dryers

- Drying is faster because inside the dryer it is warmer than outside.
- Less risk of spoilage because of the speed of drying. (if the drying process is slow the fruit start to ferment and the product is spoilt).
- The product is protected against flies, pests, rain and dust.
- It is labour saving. The product can be left in the dryer overnight or during rain.
- The quality of the product is better in terms of nutrients, hygiene and colour.

b) Vacuum drying

A drying method that places the object to be dried in an enclosed container to vent air and reduce the pressure with a vacuum pump in order to artificially increase the water vapour partial pressure difference is called vacuum drying. The vacuum drying method dries an object with a temperature of 25°C three times faster comparing to drying with warm air of 30°C and humidity of 50%. As shown in Figure 8, vacuum drying is completed in an enclosed container so that it has no benefit from heat supplied by the sun or warm air as in warm air drying. The method requires some type of heating in order to prevent the temperature drop of the drying object accompanying evaporation of moisture.

In order to prevent the temperature drop of drying object, the following conditions must take place:

- Raising the temperature prior to placing an object in a container.
- Heating the entire container to warm the drying object with radiant heat.
- Heating the drying object with conduction heat from a heater plate in the container.

The vacuum drying forces the pressure in the narrow gaps and in the tubes to decrease, which enables the moisture in the gaps to evaporate faster. In addition, moisture trapped in the narrow gaps by a bumping phenomenon could blow out, unless the temperature inside the drying object drops excessively, which further expedites drying. Therefore, using a pump with a volume sufficient for venting evaporated moisture for vacuum drying enables any porous or even powder like object to be dried thoroughly even from the inside uniformly.

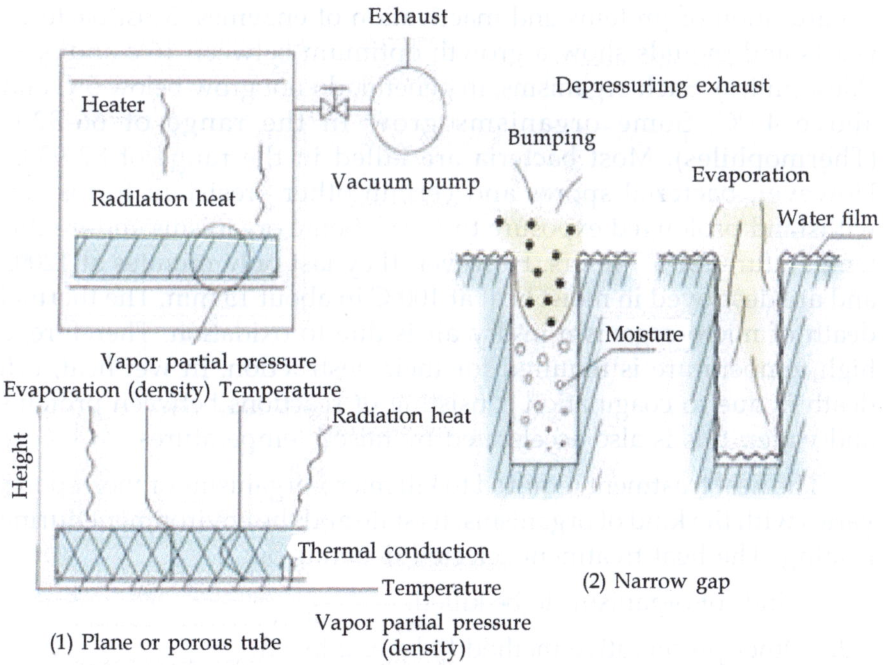

Fig. 8: Vacuum drying

However, while it has the advantages of uniform drying even from the inside of a porous object or of more rapid drying than warm air drying, the speed of drying is determined by the temperature at which a calorie removed by evaporation and a calorie supplied is balanced. In the case of porous or powder-like objects with low thermal conductivity, while the surface temperature increases from the heating method, the internal temperature does not rise, which delays the drying in the center. Thus, drying porous material with low thermal conductivity may result in completed drying of the surface by the vacuum drying method but failure to completely dry the inside.

c) Freeze drying

In this process of dehydration, the food to be dried is first frozen in a cabinet. A vacuum is created by pumping out the air in the cabinet and by the process of sublimation, ice turns into vapour. Freeze-dried food needs no further refrigeration or preservation. It is light in weight. It is used to dry fruits, vegetables, meat, poultry, and seafood.

Heat preservation: This is very commonly used method of food preservation. The micro-organisms are killed by heat due to denaturation of proteins and inactivation of enzymes. Most bacteria, yeasts and moulds show a growth optimum between 16°c and 38°C (Mesophiles). Such organisms, in general, do not grow below 5°C and above 45⁰C. Some organisms grow in the range of 66-82°C (Thermophiles). Most bacteria are killed in the range of 82-93°C. However, bacterial spores and certain other –resistant forms can withstand prolonged exposure to 100°C. Some organisms endure this temperature for $5^{1/2}$ hours. However, they last only minutes at 120°C and are destroyed in moist heat at 100°C in about 15 min. The thermal death of micro-organism in dry air is due to oxidation. Therefore, a high temperature is required for their destruction. In wet heat, cell death is due to coagulation consisting of reactions between proteins and water; this is also accelerated by raised temperatures.

The heat treatment required to kill micro-organisms or their spores varies with the kind of organisms, its state and the environment during heating. The heat treatment selected depends on:

1. Kinds of organisms to be killed.
2. Other preservative method to be employed.
3. Nature of food to be preserved.
4. Effect of heat on the food.

Food concentration

Foods are concentrated for the same reason that they are dried-preservation and reduction in weight and bulk. Relatively a few liquid foods are preserved by concentration, mainly because of the reduction in water activity (aw) and development of osmotic pressure which retard the microbial growth and enzymatic reactions. Concentration of food is usually done for reasons like: Reduction in volume and weight, reduction in packaging, storage and transport costs, better microbial stability and convenience. Examples of food preserved by concentration are tomato paste, fruit juice concentrate, soup and condensed milk. The main requirement to improve processing of these products is to control the rate of heating become thickened towards the end of boiling.

Advantages of drying

1. Dried food is easy to transport and store.
2. It has a long shelf-life provided it remains dry.
3. It occupies less storage space.
4. It is a cheap method of preservation and is convenient to use.

5. Food irradiation

Food irradiation is another sterilizing technique in which the foods are bombarded by high-energy rays, (Wavelengths<2A) or by fast moving electrons emitted from excited radioactive elements and accelerated negatively charged particles, electrons (Beta-particles), which are emitted from a hot cathode. Gamma rays are very similar to X-rays, except that they have much greater penetrating power. An electron beam can be accelerated to very high speed and increased energies so that it can penetrate foods, by passing it through electronic devices.

Low dose: Upto 1 kGy designed to control insects in grains and inhibit sprouting in white potatoes and decay and control of insects in fruits and vegetables.

Medium dose: 1 to 10 kGy designed to delay mould growth on strawberries and other fruits.

High dose: Greater than 10 kGy designed to kill micro-organisms and insects in spices and to commercially sterilize food, destroying all microbes of public health concern.

Radurization : Low doses (under 1 kGy) – inhibits the sprouting of produce (Onion, potato and garlic); retards the ripening and fungi deterioration of the fruits and vegetables (strawberry, tomato), and promotes insect disinfestations in cereals and vegetables.

Rededication (Pasteurization): Typical doses (1 to 10 kGy) – controls the presence of pathogenic organisms, especially in fruit juices; retards the deterioration of fishes and fresh meat, and controls Salmonella in poultry products.

Radapertization (Industrial Sterilization): High doses (over 10 kGy)- little used in food processing is rather important to the sterilization of health and personal hygiene products.

Applications

Ultraviolet radiations are also employed in preservations, but they have a very low degree of penetration and are employed to inactivate micro-organisms on the surface of food and to treat air, water and the surface of food processing equipment to reduce the number of micro-organisms cobalt-60 or cesium-137 or electron-producing machines are the principal sources of ionizing radiations used for food irradiation. The unit of radiation is in terms of rads (kilorad or megarad). UV irradiation is being applied commercially in the following processes of food:

- Tenderizing or aging of meat
- Curing and wrapping of cheese.
- Prevention of surface mould growth on bakery products.
- Air purification in bottling and food processing.
- Treatment of knives for slicing bread.
- Treatment of bread and cakes.
- Sanitizing of utensils used for eating.
- Prevention of growth of film of yeasts on pickles, vinegar.
- Killing of spores on sugar crystals and in syrup.
- Treatment of air in storage and processing rooms.

Effects of radiations on nutrients

- Irradiation produce molecular changes in starch, converting it into sugar. It produces denaturation and coagulation of milk protein, e.g. casein.

- Lossess of amino acids also occur ranging from 5 to 7% e.g. tryptophan, methionine, phenylalanine, whereas there is an increase upto 60% in valine content of irradiated foods.
- Lipids are very sensitive to radiations. Auto-oxidation of lipids increases peroxide value and develops rancidity.
- During irradiation 2/3rd of vitamin B_1 is found to be lost and only 37% of B_1 is retained.
- Frozen foods if irradiated, then vitamin C losses are very less.
- Vitamin K is sensitive to radiation and is destroyed in appreciable quantity during irradiations.

Uses of Radiations

Used for sterilization of foods in hermetically sealed packs.

Poultry products, seafoods and various fruits, such as apples, mangoes, banana, etc. can be irradiated and preserved at low temperature for a long period.

Prevents sprouting of potatoes, onions, etc. helps in long storage are of great economic value.

It helps in delayed ripening in fruits, therefore, exports of fruits to foreign countries without deterioration in quality is done.

Prevention of insect infestation in dry foods and food products- ionizing radiations are lethal to different stages of insects and also to eggs. Nuts, dried fruits and processed cereal foods can be preserved for a long period after disinfestations by radiation treatment.

6. Application of Fermentation in Food Preservation

Fermentation is one of the oldest method of preservation. Any other methods of preservation, such as drying, salting, sugaring, pickling, *etc.* have the common objective of decreasing the number of microorganisms in food or holding them against further multiplication. In fermentation, multiplication of micro-organisms and their metabolic activities are encouraged. However, the organisms that are encouraged to grow and multiply are the select groups whose metabolic products help food preservation.

The term 'fermentation' means the breakdown of carbohydrate material by micro-organisms (or enzymes) under anaerobic conditions.

In the fermentation of foods, a complex mixture of carbohydrates, proteins, fats, *etc.* undergo modifications simultaneously under the action of a variety of micro-organisms and enzymes present. Thus, the carbohydrates and carbohydrate-like materials undergo fermentation, proteinaceous material undergo proteolysis or "putrefactive" breakdown and lipids "lipolytic" breakdown. The nature and extent of these changes depend upon the food, types of micro-organisms present and conditions affecting their growth and metabolic pattern. The preservative effect of fermentation is caused by the chemicals excreted by the micro-organisms. The principal chemicals involved are acids (especially lactic acid) and alcohol. These inhibit the growth of common pathogenic organisms in foods.

The list of foods produced by fermentation is extremely long and includes the following: Cheese, curd, butter, all alcoholic beverages, pickles, sauerkraut, vinegar, bread, idli, soya sauce, coffee, tea and cocoa.

The benefits of fermentation are :

1. It produces flavour and textural changes.
2. The products are more nutritious than unfermented, synthesize several vitamins and other growth factors.

❑❑❑

Chapter 5

Introduction to Food and Nutrition

Definitions

a) **Food:** Foods are substances we eat daily to fulfill needs for providing energy (work), growth/tissue repair and regulating body processes (maintenance) of the human body.

b) **Nutrients:** Food contains certain substances called nutrients which are used in the normal metabolism. In the body nutrients are responsible for the function of body. The six main types of nutrients are carbohydrates, proteins, fats, minerals, vitamins, water.

c) **Nutrition**: Nutrition is a basic human need and a prerequisite to a healthy life. A proper diet is essential from the very early stages of life for proper growth, development and to remain active. Food consumption, which largely depends on production and distribution, determines health and nutrition of the population. Nutrition is the study of food in relation to health and the processes by which living organisms use food for the maintenance of life, growth, normal functioning of body organs and tissues and the production of energy.

d) **Health:** Health is not just a mere absence of disease but it is a state of complete physical, mental and social well-being.

e) **Nutritional status:** The condition of the body resulting from intake, absorption, and use of food is called nutritional status. A diet which meets body's nutritional needs generally leads to a satisfactory nutritional status.

f) **Malnutrition :** The condition resulting from either prolonged undernutrition or overnutrition is termed malnutrition. In developing countries, the main type of malnutrition is undernutrition.

g) **Overnutrition:** The condition resulting from any one food is being eaten over a considerable period of time than the requirement and necessary to meet body's demands produces a state of overnutrition.

h) **Undernutrition:** If a diet fails to meet nutritional demands of the body over a considerable period of time, that result is undernutrition.

i) **Body mass index(BMI):** It is the estimate of body fatness and health risk.

J) **Calorie or Kilocalorie**: It is defined as the quantity of heat required to raise the temperature of 1 kg of water through 1^0C.

k) **Joule (J)** : The international unit for energy is the joule (J) and it is the energy expended, when 1 kilogram (kg) is moved 1 metre (m) by a force of 1 newton (N).

l) **Basal metabolism :** The energy metabolism of a subject at complete physical and mental rest and having normal body temperature and in the post-absorptive state(*i.e* 12 hours after the intake of last meal) is known as Basal metabolism.

m) **Physiological fuel value :** The physiological fuel value of a food or a food component may be determined by measuring the heat of combustion of the food in a calorimeter and then multiplying the heat of combustion by correction factors for incomplete digestion and incomplete oxidation of the food in the body. In about 1900, Wilbur Olin Atwater and his associates at the Connecticut (Storrs) Agriculture Experiment Station used this approach to determine the physiological fuel values of a number of food components (*i.e.*, the protein, fat, and carbohydrate isolated from various foods). They determined factors appropriate for individual foods or groups of foods, and they proposed the general physiological fuel equivalents of 4.0, 8.9, and 4.0 kcal per gram of dietary protein, fat, and carbohydrate respectively for application to the mixed American diet. These factors are commonly rounded to 4, 9, and 4 kcal per gram (17, 36, and 17 kJ per gram) respectively for protein, fat, and carbohydrate. The conversion factors determined by Atwater and his associates remain in use in the twenty-first century, and energy values of foods are calculated using these factors. The energy values (physiological fuel values) reported in food composition tables are commonly estimated by determination of the proximate composition of each food (*i.e.*, the water, protein, fat, carbohydrate, and ash contents) followed by multiplication of the amount of each energy-yielding component by the appropriate conversion factor.

Chapter **6**

Nutritional Disorders

The Spectrum of Malnutrition

During the last 30 years major advances have occurred in food production, largely as a result of the adoption of 'green revolution' technology and the development of high-yielding varieties of cereals. As a result of this we live today in a world of abundance compared with the relatively recent past. The proportion of population in the developing world that is malnourished fell from 46.5% in the early 1960s to 26.7% in 2000. Despite these improvements, hunger and malnutrition continue to dominate the health of the world's poorest nations. In addition to the problem of insufficient protein and energy intake, deficiencies of micronutrients such as iron, iodine, and vitamin A affect millions of people in the developing world. Different nutrient deficiency disorders are being listed in Table 17.

Table 17: List of Deficiency Diseases

Vitamins	*Deficiency diseases*
Vitamin A or Carotene	(1) Phyrnoderma – toad skin (2) Night blindness (3) Xerophthalmia – dryness of conjunctiva of the eyes (4) Keratomalacia – ground glass appearance of the eye (5) Defective teeth-lack of enamel (6) Reduced resistance to infections (7) Tendency to the formation of kidney stones.
Vitamin D	(1) Rickets in children (2) Osteomalacia in adults-generally in women who observe purdah and remain indoors.
Vitamin B (Thiamine)	Dry beriberi- (1) Loss of appetite (2) Numbness of the legs and hands (3) Paralysis of the legs and hands (4) Drooping footsWet beriberi- (1) Oedema of legs and abdomen due to

	accumulation of fluid (2) Enlargement of the heart and signs of heart failure (3) Palpitation and breathlessness.
Riboflavin	(1) Burning sensation in the eyes, lips and tongue (2) Inflammation of the cornea of the eyes (3) Dimness of vision (4) Inflammation of the tongue (5) Dermatitis—folds behind the ears, skin, etc.
Niacin or Nicotinic Acid	Pellagra— (a) Dark pigmentation of exposed parts of the body (b) Soreness of mouth and tongue (c) Diarrhoea (d) Mental disturbances (e) Anaemia
Vitamin C or Ascorbic acid	Scurvy—(1) Weakness and irritability (2) Bleeding of skin and joints (3) Tender, swollen, bleeding gums (4) Delayed healing of wounds (5) Anaemia (6) Weak bones which easily fracture.
Vitamin E	Lessens the fertility in the men and women. In human beings vitamin E deficiency is not frequently reported.
Vitamin K	Leads to generalised bleeding with an increased tendency to haemorrhage.

Protein-energy malnutrition, its implications on health and socio-economic status of the affected individual

1. Protein-Energy Malnutrition (PEM)

It is one of the largest public health problems of the country. As the name suggests, this condition is a deficiency of protein and calories in the diet. Strictly speaking, it is not one disease, but a spectrum of conditions arising from an inadequate diet. Although it affects people of all ages, the results are most dramatic in childhood due to the highest requirements in that period. It is a condition in India, due primarily due to deficiency of total dietary calories; the protein deficiency being only secondary.

Causes and epidemiology

Factors involved in the etiology of PEM are:

- The young child's high needs for both energy and protein per kilogram relative to those of older family members;
- Inappropriate weaning practices;
- Inappropriate use of infant formula in place of breastfeeding for very young infants in poor families;
- Staple diets that are often of low energy density (not infrequently bulky and unappetizing), low in protein and fat content and not fed frequently enough to children;
- Inadequate or inappropriate child care because of, for example, time constraints for the mother or lack of knowledge regarding the importance of exclusive breastfeeding;
- Inadequate availability of food for the family because of poverty, inequity or lack of sufficient land, and problems related to intrafamily food distribution;
- Infections (viral, bacterial and parasitic) which may cause anorexia, reduce food intake, hinder nutrient absorption and utilization or result in nutrient losses;
- Famine resulting from droughts, natural disasters, wars, civil disturbances, etc.

The percentage of children classified as having severe, moderate and mild PEM depends on how these terms are defined. The two severe forms of malnutrition, kwashiorkor and nutritional marasmus, have very different appearances and clinical features as described below. It is generally agreed that the hallmark of kwashiorkor is pitting oedema, and the overriding feature of nutritional marasmus is severe underweight. Children who have both oedema and severe underweight are diagnosed as having marasmic kwashiorkor.

Implications of PEM

They can be of two types. viz., (i) medical, and (ii) socio-economic as follows:

Medical implications

- High morbidity, increased infection rates.
- Blindness.
- High morality due to otherwise controllable infections.
- Stunted growth.
- Poor energy metabolism.
- Reduces intestinal function leading to incomplete digestion and assimilation of food.
- Danger of chromosomal abnormalities.
- Effect on brain development.
- Reduced mental and psychomotor development.

Socio-economic implications

- Wastage of children and other inputs therein.
- Wastage of working hours.
- Stimulus to produce more children leading to flopping of family planning programme.
- They grow to become voracious eaters leading to increased food demands.
- Reduced working capacity causing reduced productivity, leading to poverty and a vicious circle.
- Other socio-economic implications.

The signs suggestive of protein -energy malnutrition in young children are:

- Oedema, dyspigmentation of the hair, easy pluckability of the hair, thin sparse hair,
- Straight hair, muscle wasting, diffuse depigmentation of the skin, psychomotor change,
- Moon-face, hepatomegaly and flaky-paint dermatosis.

The clinical picture varies greatly in protein- calorie malnutrition in the early years of life, not only between the two major severe syndromes, Kwashiorkor and nutritional marasmus, but with regard to the lesser and much commoner degrees of deficiency.

Anthropometric measurement. Growth retardation, as shown by a low body weight for age and by muscle depletion are characteristic findings.

Biochemical tests. Hypo-albuminaemia is present in advanced cases, and a low urinary creatinine excretion may be found. Poor dietary protein intake is reflected by a low excretion of urea per gram of creatinine. The serum amino-acid imbalance test and the hydroxyproline excretion test may show abnormality.

Caloric over nutrition in children and adults (obesity)

The signs of caloric over nutrition are:

Increased subcutaneous fat and increased abdominal girth.

Three categories of mild to moderate PEM based on weight and height measurements of children. Subsequently these categories came to be known as follows:

- **Wasting**: acute current, short-duration malnutrition, where weight for age and weight for height are low but height for age is normal;
- **Stunting**: past chronic malnutrition, where weight for age and height for age are low but weight for height is normal;
- **Wasting and stunting**: acute and chronic or current long-duration malnutrition, where weight for age, height for age and weight for height are all low.

The degree of malnutrition is always based on the child's percentage of standard weight for age. In many countries the Gomez classification is very widely used (Table 18).

Table 18: Gomez classification of malnutrition based on weight-for-age standards

Classification	Percentage of standard weight for age
Normal	>90
Grade I (mild malnutrition)	75-89.9
Grade II (moderate malnutrition)	60-74.9
Grade III (severe malnutrition)	<60

a) Kwashiorkor

Kwashiorkor is one of the serious forms of PEM. It is seen most frequently in children one to three years of age, but it may occur at any age. It is found in children who have a diet that is usually insufficient in energy and protein and often in other nutrients. Often the food provided to the child is mainly carbohydrate; it may be very bulky, and it may not be provided very frequently.

Kwashiorkor is often associated with, or even precipitated by, infectious diseases. Diarrhoea, respiratory infections, measles, whooping cough, intestinal parasites and other infections are common underlying causes of PEM and may precipitate children into either kwashiorkor or nutritional marasmus. These infections often result in loss of appetite, which is important as a cause of serious PEM. Infections, especially those resulting in fever, lead to an increased loss of nitrogen from the body which can only be replaced by protein in the diet.

Clinical signs of kwashiorkor

Kwashiorkor is relatively easy to diagnose based on the child's history, the symptoms reported and the clinical signs observed.

All cases of kwashiorkor have oedema to some degree, poor growth, wasting of muscles and fatty infiltration of the liver. Other signs include mental changes, abnormal hair, a typical dermatosis, anaemia, diarrhoea and often evidence of other micronutrient deficiencies.

Oedema. The accumulation of fluid in the tissues causes swelling; in kwashiorkor this condition is always present to some degree. It usually starts with a slight swelling of the feet and often spreads up the legs. Later, the hands and face may also swell. To diagnose the presence of oedema, the medical attendant presses with a finger or thumb above the ankle. If oedema is present the pit formed takes a few seconds to return to the level of the surrounding skin.

Poor growth. Growth failure always occurs. If the child's precise age is known, the child will be found to be shorter than normal and, except in cases of gross oedema, lighter in weight than normal (usually 60 to 80 per cent of standard or below 2 SD). These signs may be obscured by oedema or ignorance of the child's age.

Wasting. Wasting of muscles is also typical but may not be evident because of oedema. The child's arms and legs are thin because of muscle wasting.

Fatty infiltration of the liver. This condition is always found in post-mortem examination of kwashiorkor cases. It may cause palpable enlargement of the liver (hepatomegaly).

Mental changes. Mental changes are common but not invariably noticed. The child is usually apathetic about his or her surroundings and irritable, when moved or disturbed. The child prefers to remain in one position and is nearly always miserable and unsmiling. Appetite is nearly always poor.

Hair changes. The hair of a normal Asian, African or Latin American child is usually dark black and coarse in texture and has a healthy sheen that reflects light. In kwashiorkor, the hair becomes silkier and thinner. African hair loses its tight curl. At the same time it lacks lustre, is dull and lifeless and may change in colour to brown or reddish brown. Sometimes small tufts can be easily and almost painlessly plucked out. On examination under a microscope, plucked hair exhibits root changes and a narrower diameter than normal hair. The tensile strength of the hair is also reduced. In Latin America bands of discoloured hair are reported as a sign of kwashiorkor. These reddish-brown stripes have been termed the "flag sign" or "signa bandera".

Skin changes. Dermatosis develops in some but not all cases of kwashiorkor. It tends to occur first in areas of friction or of pressure, such as the groin, behind the knees and at the elbow. Darkly pigmented patches appear, which may peel off or desquamate. The similarity of these patches to old sun-baked, blistered paint has given rise to the term "flaky-paint dermatosis". Underneath the flaking skin are atrophic depigmented areas which may resemble a healing burn.

Anaemia. Most cases have some degree of anaemia because of lack of the protein required to synthesize blood cells. Anaemia may be complicated by iron deficiency, malaria, hookworm, etc.

Diarrhoea. Stools are frequently loose and contain undigested particles of food. Sometimes they have an offensive smell or are watery or tinged with blood.

Moonface. The cheeks may appear to be swollen with either fatty tissue or fluid, giving the characteristic appearance known as "moonface".

Signs of other deficiencies. In kwashiorkor some subcutaneous fat is usually palpable, and the amount gives an indication of the degree of energy deficiency. Mouth and lip changes characteristic of vitamin B deficiency are common. Xerosis or xerophthalmia resulting from vitamin A deficiency may be seen. Deficiencies of zinc and other micronutrients may occur. Refer to Fig. 8.which illustrates the clinical forms of PEM.

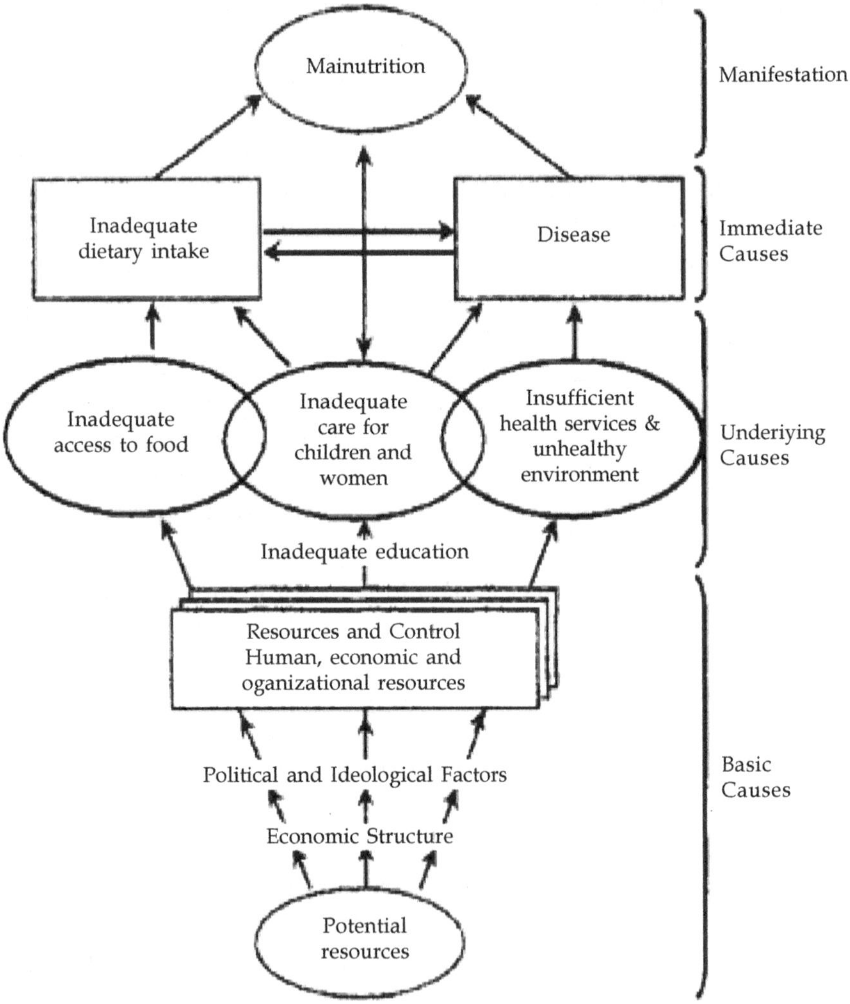

Fig. 9: Occurrence of Malnutrition

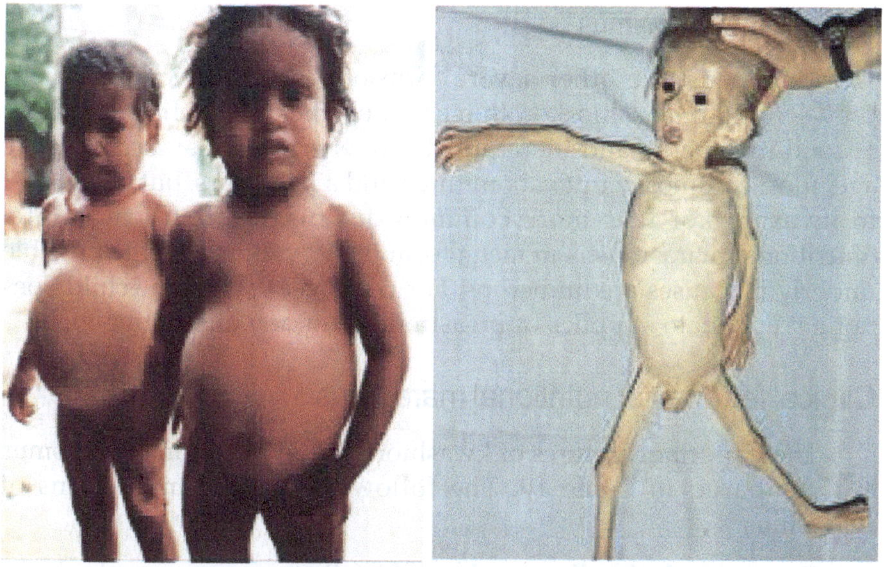

(b) A typical case of Kwashiorkor (c) A Marasmic child

Fig. 10: Clinical forms of PEM

Table 19: Comparison of the features of Kwashiorkor and Marasmus

Feature	Kwashiorkor	Marasmus
Growth failure	Present	Present
Wasting	Present	Present, marked
Oedema	Present (sometimes mild)	Absent
Hair changes	Common	Less common
Mental changes	Very common	Uncommon
Dermatosis, flaky-paint	Common	Does not occur
Appetite	Poor	Good
Anaemia	Severe (sometimes)	Present, less severe
Subcutaneous fat	Reduced but present	Absent
Face	May be oedematous	Drawn in, monkey-like
Fatty infiltration of liver	Present	Absent

b) Nutritional marasmus

Marasmus, the other severe form of PEM, is now much more prevalent than kwashiorkor. In marasmus, the main deficiency is one of food in general, and therefore also of energy. It may occur at any age, most commonly upto about three and a half years, but in contrast to kwashiorkor it is more common during the first year of life. Nutritional marasmus is in fact a form of starvation, and the possible underlying causes are numerous. For whatever reason, the child does not get adequate supplies of breastmilk or of any alternative food.

Clinical features of nutritional marasmus

The important features of kwashiorkor and nutritional marasmus are compared in Table 19. The following are the main signs of marasmus:

Poor growth. In all cases the child fails to grow properly. If the age is known, the weight will be found to be extremely low by normal standards (below 60 percent or -3 SD of the standard). In severe cases, the loss of flesh is obvious: the ribs are prominent; the belly, in contrast to the rest of the body, may be protuberant; the face has a characteristic simian (monkey-like) appearance; and the limbs are very emaciated. The child appears to be skin and bones. An advanced case of the disease is unmistakable, and once seen is never forgotten.

Wasting. The muscles are always extremely wasted. There is little if any subcutaneous fat left. The skin hangs in wrinkles, especially around the buttocks and thighs. When the skin is taken between forefinger and thumb, the usual layer of adipose tissue is found to be absent.

Alertness. Children with marasmus are quite often not disinterested like those with kwashiorkor. Instead the deep sunken eyes have a rather wide-awake appearance. Similarly, the child may be less miserable and less irritable.

Appetite. The child often has a good appetite. In fact, like any starving being, the child may be ravenous. Children with marasmus often violently suck their hands or clothing or anything else available. Sometimes they make sucking noises.

Anorexia. Some children are anorexic.

Diarrhoea. Stools may be loose, but this is not a constant feature of the disease. Diarrhoea of an infective nature, as mentioned above, may commonly have been a precipitating factor.

Anaemia. Anaemia is usually present.

Skin sores. There may be pressure sores, but these are usually over bony prominences, not in areas of friction. In contrast to kwashiorkor, there is no oedema and no flaky-paint dermatosis in marasmus.

Hair changes. Changes similar to those in kwashiorkor can occur. There is more frequently a change of texture than of colour.

Dehydration. Although not a feature of the disease itself, dehydration is a frequent accompaniment of the disease; it results from severe diarrhoea (and sometimes vomiting).

c) Marasmic Kwashiorkor

Children with features of both nutritional marasmus and kwashiorkor are diagnosed as having marasmic kwashiorkor. In the Wellcome classification (see above) this diagnosis is given for a child with severe malnutrition who is found to have both oedema and a weight for age below 60 per cent of that expected for his or her age. Children with marasmic kwashiorkor have all the features of nutritional marasmus including severe wasting, lack of subcutaneous fat and poor growth, and in addition to oedema, which is always present, they may also have any of the features of kwashiorkor described above. There may be skin changes including flaky-paint dermatosis, hair changes, mental changes and hepatomegaly. Many of these children have diarrhoea.

Treatment of severe PEM

Diet. Treatment is often based on dried Protein based diets. Mainly protein and energy based diet are advised.

Rehydration. Children with kwashiorkor or nutritional marasmus who have severe diarrhoea or diarrhoea with vomiting may be dehydrated. Intravenous feeding is not necessary unless the vomiting is severe or the child refuses to take fluids orally. Rehydration should be achieved using standard oral rehydration solution (ORS), as is described for the treatment of diarrhea. If standard ORS packets are used which are normally added to 1 litre of boiled water, in a serious case a packet might be added to 1.5 litres of water.

Medication. Although it is useful to establish standard procedures for treating kwashiorkor and nutritional marasmus in any hospital or other health unit, each case should nevertheless be treated on its own merits.

2. Iron deficiency and other nutritional anaemias

Nutritional anaemias are extremely prevalent worldwide. Unlike protein-energy malnutrition (PEM), vitamin A deficiency and iodine deficiency disorders (IDD), these anaemias occur frequently in both developing and industrialized countries. The most common cause of anaemia is a deficiency of iron, although not necessarily a dietary deficiency of total iron intake. Deficiencies of folates (or folic acid), vitamin B_{12} and protein may also cause anaemia. Ascorbic acid, vitamin E, copper and pyridoxine are also needed for production of red blood cells (erythrocytes). Vitamin A deficiency is also associated with anaemia.

Classification of Anaemia

Some anemia are classified as sickle cell anaemia, aplastic anaemias, thalassaemias and severe haemorrhage, and based on the characteristics of the blood cells or other features, anaemias may be classified as microcytic (having small red blood cells), macrocytic (having large red blood cells), haemolytic (having many ruptured red blood cells) or hypochromic (having pale-coloured cells with less haemoglobin). Macrocytic anaemias are often caused by folate or vitamin B_{12} deficiencies.

In anaemia, the blood has less haemoglobin than normal. Haemoglobin is the pigment in red cells that gives blood its red colour. It is made of protein with iron linked to it. Haemoglobin carries oxygen in the blood to all parts of the body. In anaemia, either the amount of haemoglobin in each red cell is low (hypochromic anaemia) or there is a reduction in the total number of red cells in the body. This process requires adequate amounts of nutrients, especially iron, other minerals, protein and vitamins, all of which originate in the food consumed. Iron deficiency is the most prevalent important nutritional problem of humans.

Although deficiency diseases are usually considered mainly as consequences of a lack of the nutrient in the diet, iron deficiency anaemia occurs frequently in people whose diets contain quantities of iron close to the recommended allowances. However, some forms of iron are absorbed better than others; certain items in the diet enhance or detract from iron absorption; and iron can be lost because of many conditions, an important one in many tropical countries being hookworm infection, which is very common.

Causes and epidemiology

To maintain good iron nutritional status each individual needs to have an adequate quantity of iron in the diet. The iron has to be in a form that permits a sufficient amount of it to be absorbed from the intestines. The absorption of iron may be enhanced or inhibited by other dietary substances.

Human beings have the ability both to store and to conserve iron, and it must also be transported properly within the body. The average male adult has 4 to 5 g of iron in his body, most of it in haemoglobin, a little in myoglobin and in enzymes and around 1 g in storage iron, mainly ferritin in the cells, especially in the liver and bone marrow. Losses of iron from the body must not deplete the supply to less than that needed for manufacture of new red blood cells.

Of the dietary deficiency causes of nutritional anaemias, iron deficiency is clearly by far the most important. Good dietary sources of iron include foods of animal origin, such as liver, red meat and blood products, all containing haem iron, and vegetable sources, such as some pulses, dark green leafy vegetables and millet, all containing non-haem iron. However, the total quantity of iron in the diet is not the only factor that influences the likelihood of developing anaemia. The type of iron in the diet, the individual's requirements for iron, iron losses and other factors often are the determining factors. Iron absorption is influenced by many factors. High iron is needed by pregnant women and growing children.

The availability of iron in foods varies greatly. In general, haem iron from foods of animal origin (meat, poultry and fish) is well absorbed, but the non-haem iron in vegetable products, including cereals such as wheat, maize and rice, is poorly absorbed. These differences may be modified when a mixture of foods is consumed. It is well known that phytates and phosphates, which are present in cereal grains, inhibit iron absorption. On the other hand, protein and ascorbic acid (vitamin C) enhance iron absorption.

Although most solid diets, both for children and adults, provide the recommended allowances for iron, the iron may be poorly absorbed. Many people have increased needs because of blood loss from hookworm or bilharzia infections, menstruation, childbirth or wounds. Women have increased needs during pregnancy, when iron is needed for the foetus, and during lactation, for the iron in breastmilk. It is stressed that iron from vegetable products, including cereal grains, is less well absorbed than that from most animal products.

Anaemia resulting from folate deficiency is less prevalent than that from iron deficiency or iron loss. It occurs when folate intakes are low and when red cells are haemolysed or destroyed in conditions like malaria. The anaemia of both folate and vitamin B_{12} is macrocytic, with larger than normal red blood cells. Folic acid or folates are present in many foods including foods of animal origin (*e.g.* liver and fish) and of vegetable origin (e.g. leafy vegetables). Vitamin B_{12} is present only in foods of animal origin.

Clinical manifestations

Haemoglobin in the red blood cells is necessary to carry oxygen, and many of the symptoms and signs of anaemia result from the reduced capacity of the blood to transport oxygen.

The symptoms and signs are:

- Tiredness, fatigue and lassitude;
- Breathlessness following even moderate exertion;
- dizziness and/or headaches;
- Palpitations, with the person complaining of being aware of his or her heartbeat;
- Pallor of the mucous membranes and beneath the nails;
- Oedema (in chronic, severe cases).

These symptoms and signs are not confined to iron deficiency anaemia but are similar in most forms of anaemia. Some other symptoms are :

- Paler nails
- Enlargement of the heart
- Oedema usually occurs first in the feet and at the ankles.
- Increased pulse rate or tachycardia.
- Occasionally the nails become relatively concave rather than convex and become brittle. Anaemia is also reported to lead both to abnormalities of the mouth such as glossitis and to pica (abnormal consumption of earth, clay or other substances).

The normal range of haemoglobin is as follows:

Haemoglobin is tested by taking a sample of blood and the normal levels are as follows:

Male	:	13.8 to 17.2 gm/dl
Female	:	12.1 to 15.1 gm/dl
Pregnant Women	:	11.0 to 12.0 gm/dl
Children	:	11.0 to 16.0 gm/dl

where gm/dl = grams/decilitre

Anaemia leads to a reduced ability to do heavy work for long periods; to slower learning and more difficulty in concentration by children in school or elsewhere; and to poorer psychological development. A very important aspect of anaemia in women is that it markedly increases the risk of death of the mother during or after childbirth.

Laboratory tests

The diagnosis of anaemia requires a laboratory test. In this respect it differs from the serious manifestations of PEM, vitamin A deficiency and IDD; kwashiorkor, nutritional marasmus, advanced xerophthalmia, goitre and cretinism can all be diagnosed with some degree of certainty. Determinations of haemoglobin or haematocrit levels are the most widely used in the diagnosis of anaemia..

Treatment and dietary intervention

The treatment of anaemia depends on the cause. Iron deficiency anaemia is relatively easy and very cheap to treat. There are many different iron preparations on the market; ferrous sulphate is among the cheapest and most effective. The recommended dose of ferrous sulphate is usually 300 mg (providing 60 mg of elemental iron) twice daily between meals for adults.

Consuming of green leafy vegetables, pulses, red meat, dry fruits along with citrus fruits will reduce the incidences of iron deficiency anaemia.

It is common during pregnancy to provide folate as well as iron, or combined with iron, as part of the treatment of or prophylaxis against anaemia. For prevention, where anaemia is prevalent, doses of 120 mg of iron and 5 mg of folate daily are recommended. For treatment of established anaemia, doses of 180 mg of iron and 10 mg of folate are suggested.

- In vitamin B_{12} deficiency an oral dose of 1 μg vitamin B_{12} daily is needed.
- In India, adolescent girls are being provided Iron tables at ICDS centres.
- Green leafy vegetable, fortified foods
- Citrus fruits and ascorbic acids are important to overcome iron deficiency anaemia.

3. Vitamin A deficiency

Vitamin A was discovered in 1913, when experiments showed that if the only fat present in diets of young animals was lard, their growth was retarded, and when butter was substituted the animals grew and thrived. A substance in butter but not in lard was found also in egg yolk and cod-liver oil. It was named vitamin A. It was later established that many products of vegetable origin had nutritional properties similar to those presented by vitamin A in foods of animal origin; they were found to contain a yellow pigment, carotene, which is converted to vitamin A in the body. Preformed vitamin A or retinol is a fat-soluble vitamin found only in animal products. Carotenes or carotenoids can act as a provitamin. There are many carotenoids in plants, but the most important for human nutrition is beta-carotene, which can be converted to vitamin A by enzymatic action in the intestinal wall. Breastmilk is an important source of vitamin A for infants.

Dietary deficiency of vitamin A most commonly and importantly affects the eyes, and it can lead to blindness. Xerophthalmia, meaning drying of the eyes (from the Greek word *xeros,* meaning dry), is the term now used to cover the eye manifestations resulting from vitamin A deficiency. Vitamin A deficiency also has a role in a variety of clinical conditions not related to the eyes, and it may contribute to higher child mortality rates, especially in children who develop measles. Diets deficient in vitamin A (very good antioxidant) have increased rates and severity of infections. Vitamin A deficiency also adversely affects epithelial surfaces apart from the eye and is associated with an increased incidence of certain cancers, including cancer of the colon. The serious eye manifestations of vitamin A deficiency leading to corneal destruction and blindness are mainly seen in young children. This condition is sometimes called keratomalacia.

Until recently vitamin A deficiency was a relatively neglected condition, however, recently more focus is paid to virtual elimination of vitamin A deficiency and its consequences, including blindness.

Causes : An inadequate intake of carotene or preformed vitamin A, poor absorption of the vitamin or an increased metabolic demand can all lead to vitamin A deficiency. Of these three, dietary deficiency is by far the most common cause of xerophthalmia.

Good sources of retinol, or preformed vitamin A, are liver, fish-liver oils, egg yolks and dairy products. The yellow colour of carotene may be masked by chlorophyll in many dark green leafy vegetables. Carotenes are present in good quantities in a wide variety of green and yellow vegetables and fruits, in yellow maize and in yellow root crops, e.g. sweet potatoes, carrots, etc. A rich source is red palm oil also being advocated.

In many tropical diets important sources are dark green leafy vegetables [e.g. amaranth, cassava and drumstick *(Moringa oleifera)* leaves], mangoes, papayas, tomatoes and sometimes local yellow pumpkins, squash and yellow maize. The biological activity of vitamin A is now usually expressed as retinol equivalents (RE) rather than in international units (IU). One RE is equal to 1 μg of retinol or 6 μg of beta-carotene. The World Health Organization (WHO) has recommended an intake of 300 RE daily for infants and 750 RE for adults.

Vitamin A, either preformed (retinol) or converted from carotene, is stored in the liver. Retinol is transported from the liver to other sites in the body by retinol binding protein (RBP), a specific carrier protein. Protein deficiency may influence vitamin A status by reducing the synthesis of RBP.

Low intake of vitamin A and carotene over an extended period is the most common cause of xerophthalmia. The condition may be influenced by other factors, however, e.g. intestinal parasitic infections, gastro-enteritis or malabsorption. Measles often precipitates xerophthalmia because it leads to lowered food intake (in which anorexia and stomatitis may be factors) and to increased metabolic demands for vitamin A. The virus may also affect the eye, aggravating lesions caused by vitamin A deficiency. PEM is also important as a cause or accompaniment of xerophthalmia.

Epidemiology

Vitamin A deficiency is the most common cause of blindness in children in many endemic areas. Xerophthalmia occurs almost entirely in children living in poverty. It is extremely rare to find cases in more affluent families, even in areas where xerophthalmia is prevalent. It is a disease related to low socio-economic status, low levels of female literacy, land shortages, inequity, poor availability of curative and preventive primary health care, high rates of infectious and parasitic diseases (often related to poor sanitation and water supplies) and grossly inadequate family food security. As with PEM, three essentials for prevention of vitamin A deficiency are adequate food security, care and health.

In recent decades, Xerophthalmia has been especially prevalent in children of poor rice-eating families in South and Southeast Asia (e.g. Bangladesh, India, Indonesia and the Phillippines).

Prevalence rates of five different signs have been recommended as criteria for judging whether xerophthalmia is a significant public health problem in a given population (Table 20). It is suggested that if the prevalence of any one sign (i.e. the percentage of children examined having the sign) in children aged six months to six years in a vulnerable population is above the cut-off, then xerophthalmia should be considered a public health problem in that population.

Clinical manifestations

Vitamin A deficiency has a plethora of clinical manifestations, ranging from xerophthalmia (practically pathognomonic) to disturbances in growth and susceptibility to severe infection (far more protean). Like other classical vitamin deficiency states (scurvy, rickets), some of the signs and symptoms of xerophthalmia were recognized long ago. The risk of VAD is increased in patients suffering from fat malabsorption, cystic fibrosis, sprue, pancreatic insufficiency, IBD, or cholestasis, as well as in persons who have undergone small-bowel bypass surgery. The risk is also increased in vegans, refugees, recent immigrants, persons with alcoholism, and toddlers and preschool children living below the poverty line. These patients should be advised to consume vitamin A.

Table 20: Prevalence criteria for determining public health significance of Vitamin A deficiency

Sign	Prevalence above (%)
Night blindness	1
Bitot's spots	0.5
Corneal xerosis/corneal ulceration/keratomalacia	0.01
Corneal scar	0.05
Plasma vitamin A <10 μg/dl	5

Source: WHO, 1982

Laboratory tests

Since vitamin A is stored in the liver, a diet deficient in vitamin A results eventually in low hepatic stores. Thus the best way to judge vitamin A nutritional status is to obtain an estimate of the level of vitamin A in the liver. This level can be measured easily only at autopsy.

Determination of serum vitamin A levels is useful for community surveys. Serum retinol levels often fall from normal levels of 30 to 50 μg per 100 ml of plasma to low or deficient levels below 20 μg per 100 ml of plasma. Children with xerophthalmia will usually have levels below 10,μg per 100 ml. Ocular manifestations of xerophthalmia seldom occur before serum vitamin A levels are deficient.

Treatment

Effective treatment depends on early diagnosis, immediate dosing with vitamin A and proper treatment of other illnesses, such as PEM, tuberculosis, infections and dehydration. Severe cases with corneal involvement should be treated as emergencies. Sometimes hours, and certainly days, may make the difference between reasonable vision and total blindness.

Treatment for children should consist of 110 mg of retinyl palmitate or 66 mg of retinyl acetate (200 000 IU of vitamin A) orally or preferably 33 mg (100 000 IU) of water-miscible vitamin A (retinyl palmitate) by intramuscular injection. These doses should be halved for infants.

Night blindness and conjunctival xerosis are completely reversible and respond quickly to treatment using oral doses of vitamin A.

Prevention

In the long term, sustainable control will be achieved by increasing the production and consumption of foods rich in vitamin A and carotene by at-risk populations.

4. Iodine Deficiency Disorder

Iodine is an essential trace element; the thyroid hormones thyroxine and triiodothyronine contain iodine. In areas where there is little iodine in the diet-typically remote inland areas, where no marine foods are eaten.

Iodine deficiency gives rise to goiter (so-called endemic goiter), as well as cretinism, which results in developmental delays and other health problems.

"According to World Health Organization, in 2007, nearly 2 billion individuals had insufficient iodine intake, a third being of school age. Thus iodine deficiency, as the single greatest preventable cause of mental retardation, is an important public-health problem."

In some such areas, this is now combated by the addition of small amounts of iodine to table salt in form of sodium iodide, sodium iodate, potassium iodide, and/orpotassium iodate—this product is known as *iodized salt.* Iodine compounds have also been added to other foodstuffs, such as flour, water and milk in areas of deficiency. Seafood is also a well known source of iodine. Thus, iodine deficiency is more common in mountainous regions of the world, where food is grown in iodine-poor soil.

Goiter

A low amount of thyroxine (one of the two thyroid hormones) in the blood, due to lack of dietary iodine to make it, gives rise to high levels of thyroid stimulating hormone (TSH), which stimulates the thyroid gland to increase many biochemical processes; the cellular growth and proliferation can result in the characteristic swelling or hyperplasia of the thyroid gland, or goiter. In mild iodine deficiency, levels of triiodiothyronine (T_3) may be elevated in the presence of low levels of levothyroxine, as the body converts more of the levothyroxine to triiodothyronine as a compensation. Some such patients may have a goiter, without an elevated TSH. The introduction of iodized salt since the early 1900s has eliminated this condition in many affluent countries.

Goiter is said to be endemic, when the prevalence in a population is $> 5\%$, and in most cases goiter can be treated with iodine supplementation. If goiter is untreated for around five years, however, iodine supplementation or thyroxine treatment may not reduce the size of the thyroid gland because the thyroid is permanently damaged.

Cretinism

Iodine deficiency is one of the leading causes of preventable mental handicaps worldwide, producing typical reductions in IQ of 10 to 15 IQ points.

Cretinism is a condition associated with iodine deficiency and goiter, commonly characterised by mental deficiency, deaf-mutism, squint, disorders of stance and gait, stunted growth and hypothyroidism.

Deficiency: In areas where there is little iodine in the diet, typically remote inland areas and semi-arid equatorial climates where no marine foods are eaten, iodine deficiency gives rise to hypothyroidism, symptoms of which are extreme fatigue, goiter, mental slowing, depression, weight gain, and low basal body temperatures.

Iodine deficiency is the leading cause of preventable mental retardation, a result which occurs primarily when babies or small children are rendered hypothyroidic by a lack of the element. The addition of iodine to table salt has largely eliminated this problem.

Risk factors

Following is a list of potential risk factors that may lead to iodine deficiency:

1. Low dietary iodine
2. Selenium deficiency
3. Pregnancy
4. Exposure to radiation
5. Increased intake/plasma levels of goitrogens, such as calcium
6. Gender (higher occurrence in women)
7. Smoking tobacco
8. Alcohol (reduced prevalence in users).

Diagnosis

Biochemical tests. No simple tests are available, nor are they required for standard goitre surveys. For research purposes and if elaborate equipment is available, the urinary iodine can be assessed and various tests for thyroid function undertaken.

Management

Iodine deficiency is treated by ingestion of iodine salts, such as found in food supplements. Mild cases may be treated by using iodized salt in daily food consumption, or drinking more milk, or eating egg yolks, and saltwater fish.

1. Obesity and Overweight

Obesity is a medical condition in which excess bodyfat has accumulated to the extent that it may have an adverse effect on health leading to reduced life expectancy and increased health problems. It is a complex multifactorial chronic disease developing from interactive influence of numerous factors, such as social, behavioral, psychological, metabolic, cellular and molecular. Overweight is a condition where the body weight is 10-20 per cent greater than the mean standard weight for age, height and sex.

Prevalence

- Worldwide obesity has nearly doubled since 1980.
- In 2008, more than 1.4 billion adults, 20 and older, were overweight. Of these over 200 million men and nearly 300 million women were obese.
- 35% of adults aged 20 and over were overweight in 2008, and 11% were obese.
- 65% of the world's population live in countries where overweight and obesity kills more people than underweight.
- More than 40 million children under the age of 5 were overweight or obese in 2012.
- Obesity is preventable.

Overweight and Obesity

Overweight and obesity are defined as abnormal or excessive fat accumulation that may impair health.

Body mass index (BMI) is a simple index of weight-for-height that is commonly used to classify overweight and obesity in adults. It is defined as a person's weight in kilograms divided by the square of his height in meters (kg/m^2).

The WHO definition is:

- BMI greater than or equal to 25 is overweight.
- BMI greater than or equal to 30 is obesity.

BMI provides the most useful population-level measure of overweight and obesity as it is the same for both sexes and for all ages of adults. However, it should be considered a rough guide because it may not correspond to the same degree of fatness in different individuals.

Facts about overweight and obesity

Overweight and obesity are leading risks for global deaths. Around 3.4 million adults die each year as a result of being overweight or obese. In addition, 44% of the diabetes burden, 23% of the ischaemic heart disease burden and between 7% and 41% of certain cancer burdens are attributable to overweight and obesity.

Overweight and obesity are linked to more deaths worldwide than underweight. For example, 65% of the world's population live in countries where overweight and obesity kill more people than underweight (this includes all high-income and most middle-income countries).

Causes obesity and overweight

The fundamental cause of obesity and overweight is an energy imbalance between calories consumed and calories expended. Globally, there has been:

- An increased intake of energy-dense foods that are high in fat; and
- An increase in physical inactivity due to the increasingly sedentary nature of many forms of work, changing modes of transportation, and increasing urbanization.

Changes in dietary and physical activity patterns are often the result of environmental and societal changes associated with development and lack of supportive policies in sectors, such as health, agriculture, transport, urban planning, environment, food processing, distribution, marketing and education.

Common health consequences of overweight and obesity

Raised BMI is a major risk factor for noncommunicable diseases, such as:

- Cardiovascular diseases (mainly heart disease and stroke), which were the leading cause of death in 2012;
- Diabetes;
- Musculo-skeletal disorders (especially osteoarthritis - a highly disabling degenerative disease of the joints);
- Some cancers (endometrial, breast, and colon).

The risk for these noncommunicable diseases increases, with the increase in BMI.

Childhood obesity is associated with a higher chance of obesity, premature death and disability in adulthood. But in addition to increased future risks, obese children experience breathing difficulties, increased risk of fractures, hypertension, early markers of cardiovascular disease, insulin resistance and psychological effects.

Facing a double burden of disease

Many low- and middle-income countries are now facing a "double burden" of disease.

- While they continue to deal with the problems of infectious disease and under-nutrition, they are experiencing a rapid upsurge in non-communicable disease risk factors, such as obesity and overweight, particularly in urban settings.
- It is not uncommon to find under-nutrition and obesity existing side-by-side within the same country, the same community and the same household.

Children in low- and middle-income are more vulnerable to inadequate pre-natal, infant and young child nutrition At the same time, they are exposed to high-fat, high-sugar, high-salt, energy-dense, micronutrient-poor foods, which tend to be lower in cost but also lower in nutrient quality. These dietary patterns in conjunction with lower levels of physical activity, result in sharp increases in childhood obesity while undernutrition issues remain unsolved.

How can overweight and obesity be reduced?

Overweight and obesity, as well as their related non communicable diseases, are largely preventable. Supportive environments and communities are fundamental in shaping people's choices, making the healthier choice of foods and regular physical activity the easiest choice (accessible, available and affordable), and therefore, preventing obesity.

At the individual level, people can:

- Limit energy intake from total fats and sugars;
- Increase consumption of fruit and vegetables, as well as legumes, whole grains and nuts;
- Engage in regular physical activity (60 minutes a day for children and 150 minutes per week for adults).
- Individual responsibility can only have its full effect where people have access to a healthy lifestyle. Therefore, at the societal level it is important to:

- Support individuals in following the recommendations above, through sustained political commitment and the collaboration of many public and private stakeholders;
- Make regular physical activity and healthier dietary choices available, affordable and easily accessible to all - especially the poorest individuals.

The food industry can play a significant role in promoting healthy diets by:

- Reducing the fat, sugar and salt content of processed foods;
- Ensuring that healthy and nutritious choices are available and affordable to all consumers;
- Practicing responsible marketing especially those aimed at children and teenagers;
- Ensuring the availability of healthy food choices and supporting regular physical activity practice in the workplace.

Nutrition programmes to combat malnutrition

The National Nutrition Policy, about which we have studied above, was formulated only in 1993. However, prior to that, since the last four decades, Government has launched a variety of nutrition intervention programmes to combat malnutrition. One of the first nutrition programmes launched by the government was the Applied Nutrition Programme (ANP), way back in 1963. Thereafter, numerous programmes have been launched. Some of these programmes are in operation and some are not. Some new programmes focusing on ensuring food security for all and employment-based programmes have also been initiated by the government. We will now study about all these programmes. We will divide these programmes into the following sections:

- Integrated Child Development Services Programme (ICDS), which remains one of the world's most unique communities based, outreach programme for early childhood care and development.
- Nutrient Deficiency Control Programmes, namely National Prophylaxis Programme for Prevention of Blindness due to

Vitamin A deficiency, National Anaemia Control Programme, National Iodine Deficiency Disorder (IDD) Control Programme.

- Food Security Programmes, namely Public Distribution System (PDS), Antodaya Anna Yojna, Annapurna Scheme, National Food for Work Programme.

1. Integrated Child Development Services (ICDS)

The Integrated Child Development Services is the world's most unique welfare programme, which holistically addresses health, nutrition and development needs of young children, adolescent girls and pregnant/nursing mothers across the life cycle. Launched by the Government of India in 1975-76, in 33 blocks, today, it has expanded to 4 3 4 8 community development blocks (to target 5652 blocks upon expansion), reaching out t o 26.85 million young children, 5.07 lakh adolescent girls and 5.3 million pregnant/ nursing mothers, through a network of over 5 lakh anganwadi centres and more than 1 lakh mahila mandals.

ICDS contributes not only to the achievement of women and child goals related to health, nutrition and early child development, but also to other physically health care goals and the goals of universal elementary education, as enunciated in the National Plan of Action for Children, 1992. Integration of services and consideration of the mother and child as one 'biological unit' are the unique features of this programme. We will look at the 1) objectives, 2) target groups, 3) programme components and, 4) implementation of' ICDS. Let us begin with the objectives.

2. Objectives of the ICDS

The ICDS scheme aims at the holistic development of children in the age group of 0-6 years, nursing and pregnant mothers belonging to the most deprived sections of the Society. The specific objectives of the ICDS are to:

1. Improve the nutritional and health status of children in the age group of 0-6 years and adolescents,
2. Lay the foundation for proper psychological, physical and social development of the child, Nutrition Policy and Programmes.

3. Public Nutrition

a) Reduce the incidence of mortality, morbidity, malnutrition and school drop-out,

b) Achieve effective coordination of policy and implementation amongst the various Departments to promote child development,

c) Enhance the capability of the mother to look after the health and nutritional needs of the child through proper nutrition and health education,

d) Let us look at type of population who receives the benefits of the programme. i.e. the target groups.

4. Target Groups

The main beneficiaries of the ICDS programme are:

- Infants
- Children 1-6 years of age
- Pregnant and lactating women
- Adolescent girls and nutrition education
- All women upto 45 years of age.

5. Programming Components

ICDS programme is a package of several services. The services offered by the programme include:

- Supplementary nutrition
- Immunization
- Periodic health check-ups, treatment of minor ailments and referral services
- Growth monitoring
- Non-formal preschool education
- Health and nutrition education
- Adolescent girls scheme
- Safe drinking water

6. Foods supplementation

a) National Prophylaxis Programme for Prevention of Blindness due to Vitamin A Deficiency

In India, milder forms of vitamin A deficiency affecting conjunctiva, like bitot spots are observed in about I-5% preschool children. According to WHO, >0.5% prevalence of bitot spot in preschool children is indicative of public health significance. It is, however, important to understand that even mild vitamin A deficiency probably increases morbidity and mortality in children, emphasizing the public health importance of this disorder. Hence, the need for the National Prophylaxis Programme for the prevention of Nutritional Blindness due to vitamin A deficiency.

Objective

The National Prophylaxis Programme for the Prevention of Nutritional Blindness due to vitamin A deficiency aims at protecting children 6 months 5 years at risk from vitamin A deficiency.

Target Group

All children, of 6 months to 5 years, particularly those living in rural, tribal and urban slum areas, are beneficiaries of the programme.

Programme Strategy

The programme focuses on two strategies:

a) Prevention of Vitamin A deficiency, and

b) Treatment of Vitamin A deficiency.

a) **Prevention of Vitamin A Deficiency :** The prevention strategy within the programme comprises a long-term and a short-term intervention. While the short-term intervention focuses on administration of mega dose of vitamin A on periodic basis, dietary improvement is the long-term ultimate solution to the problem of vitamin A deficiency.

i) Long- term Intervention-Promoting consumption of vitamin A rich food

Action points under this intervention include that:

- Regular dietary intake of vitamin A rich food by pregnant and lactating mothers and by children under 5 years of age must be promoted.
- The mothers attending antenatal clinics and immunization sessions, as well as, mothers and children enrolled in the ICDS Programme must be made aware of the importance of preventing vitamin A deficiency.
- Breastfeeding including feeding of colostrum must be encouraged.
- Feeding of locally available β-carotene (precursor of vitamin A) rich food such as green leafy vegetables and yellow and orange vegetables and fruits like pumpkin, carrots, papaya, mango, oranges, etc. along with cereals and pulses to a weaning child must be promoted widely. In addition, whenever, economically feasible, consumption of milk, cheese, panel, yoghurt, ghee, eggs, liver, etc. must be promoted.

For increasing availability of vitamin A rich food, growing of vitamin A rich food in home gardens and consumption of these must be promoted.

ii) Short term intervention- massive dose of vitamin A

Administration of massive dose of vitamin A to preschool children at periodic intervals is a simple, effective and most direct intervention strategy. This is a short-term strategy. Unlike most other micronutrients, vitamin A is stored in the body for prolonged periods and hence, periodic administration of massive dose ensures adequate vitamin A. nutrition under the massive dose programme, every infant 6-11 months and children

- Five years is to be administered vitamin A every 6 months. Priority is to be given for coverage of children 6 months to 3 years since the highest prevalence of clinical signs of vitamin A deficiency is reported in this age group. The recommended schedule is as follows:

  6-11 months - one dose of 100000 IU

  1-5 years - 200 000 IU 6 months

A child must receive a total of 9 oral doses of vitamin A by the fifth birthday. The contact with an infant during administration of measles vaccine between the age of 9-12 months is considered a practical time: for administering the vitamin A supplement.

Implementation Strategy

The national prophylaxis programme for the prevention of nutritional blindness due to vitamin A deficiency is implemented through the Primary Health Centers and is subcenters. The services of ICDS, under the Department of Women and Child Development, Ministry of Welfare, is utilized for the distribution of vitamin A to children in the ICDS Blocks and for the education of mothers on prevention of vitamin A deficiency.

National Nutritional Anaemia Control Programme

Although anaemia is widespread in the country, it especially affects women in the reproductive age group and young children. It is estimated that over 50 per cent of pregnant women are anaemic. Nutritional anaemia, due to iron and folic acid deficiency, is directly or indirectly responsible for about 20 per cent of maternal deaths. Anaemia is a major contributor cause of high incidence of premature births, low birth weight and perinatal mortality. To reduce the prevalence of anaemia in pregnancy the national anaemia prophylaxis programme of iron and folic acid distribution to pregnant mothers was initiated by Government of India in 1972.

Objectives

The National Nutritional Anaemia Control Programme aims at significantly decreasing the prevalence and incidence of anaemia in women in reproductive age group, especially pregnant and lactating women, and preschool children.

Target Group

The beneficiaries of the programme include:

- Women in the reproductive age group, particularly pregnant and lactating mothers.
- Children 1-5 years of age.

Programme Strategy

The programme focuses on primarily three strategies:

a) Promotion of regular consumption of foods rich in iron

b) Promotion of consumption of iron and folic acid supplements to the 'high risk' groups, and

c) Identification and treatment of severely anaemic cases.

Promotion of regular consumption of foods rich in iron

Various action points under this strategy include :

- Regular dietary intake of iron and folic acid rich foods by pregnant and lactating women, adolescent girls and children under 5 years of age must be promoted.
- Regular consumption of iron rich foods, such as green leafy vegetables (such as mustard leaves amaranth colocasia leaves, knoll khol greens bengal gram greens turnip greens cereals (wheat, ragi, jowar, bajra), pulses, especially sprouted pulses and jaggery (gur) must be promoted widely. In addition, wherever culturally and economically feasible, consumption of animal foods such as meat, liver, poultry, etc: must be encouraged.
- Ensure incorporation of iron rich foods, such as green leafy vegetables in the weaning foods of infants.
- Vitamin C (ascorbic acid) promotes absorption of iron. Regular consumption of vitamin C rich foods, such as lemon, orange, guava, aonla, green mango along with iron rich foods must be promoted.
- For increasing availability of iron rich foods, growing of iron rich foods in home gardens and consumption of these must be promoted.
- Tea inhibits absorption of iron. Advice a reduce consumption of tea, especially during pregnancy, for improving the absorption of iron and prevention of anaemia.

Promoting consumption of iron and folic acid supplements to the 'high risk' groups

As a priority, all pregnant women, ir-respective of haemoglobin levels, must be provided with the recommended dose of iron and folic acid (folifer) supplements. Preschool -children, especially those in tribal areas and ICDS blocks, should be given on priority the recommended dosage of iron and folic acid supplements. The recommended doses of folic acid and iron supplements are:

i) Pregnant women: One big (adult) tablet per day for 100 days (each tablet containing 60 mg 100 mg of elemental iron and 500 μg folic acid). These tablets provided to women after the first trimester of pregnancy.

ii) Preschool: Children (1-5 years): One pediatric (small) tablet containing 20 mg iron and 100 μg folic acid daily for 100 days every year.

iii) Lactating women and Intrauterine device (IUD) acceptors: One big (adult) tablet (containing 60 mg/100 mg of elemental iron and 500 μg folic acid) per day for 700 days.

Programme implementation

The Multipurpose Worker (Female) and other paramedical working in the Primary Health Centers are responsible for the distribution of iron tablets (adult and pediatric doses) to the beneficiaries. Department of Food (Ministry of Food and Civil Supplies) is responsible for promoting consumption of iron rich foods. In addition, services of other community level workers and involvement formal and non-formal education, media, horticulture departments and voluntary organizations is utilized for the effective implementation of the programme. In addition, records of fewer than fives and antenatal care maintained under the MCH services and ICDS programme, is used for identifying beneficiaries, as well as, for recording and monitoring the distribution of iron and folic acid supplements.

National Iodine Deficiency Disorders Control Programme (NIDDCP)

Iodine Deficiency Disorders (IDD) form a spectrum of abnormalities which include goitre, mental retardation, deaf mutism, and squint, difficulties in standing or walking normally and stunting of the limbs. Iodine deficient women frequently suffer abortions and still births. Their children may be born deformed, intently deficient or even cretins. All these problems are caused by simple lack of iodine, and goitre is the least tragic of them. No State in India is free from IDD.

Objectives

The objectives of the NIDDCP include:

- Supply of iodized salt in place of common salt to the entire country, The emphasis is on establishing iodized salt with active private sector participation.
- Re-survey to assess the impact of supply of iodized salt.

Target Group

Entire population, particularly women in child bearing age and young children.

Implementation strategy

The NIDDCP is a executed by a multiplicity of agencies comprising the health, industry and railway ministries of the Central Government.

National Programme of Nutritional Support to Primary Education (Mid Day Meal Programme)

The National Programme of Nutritional Support to Primary Education commonly known as Mid Day Meals Scheme was launched in August, 1995 consequent to the favourable impact of the scheme on children in some States, as well as, the comfortable food stock position in the country, and to relate primary education with nutrition, health and ICDS. However, children, particularly from poor families, do not get enough food to eat. Their home diets are often inadequate. Many, especially in rural areas, come to school partly human some even on empty stomach, trekking long distances. Under such circumstances, they are unable to concentrate on the studies and benefit from the Education. Hence, providing a supplement in school would complement the home diet and sustain the interest of children in learning, so those drop-out rates are lowered and scion attendance improves.

Objectives

The programme is intended to give a boost to universalization of primary education by increasing enrolment, retention and attendance and simultaneously impacting upon nutritional status of students in primary classes.

Programme Implementation

The programme is operated by the Department of Education. The programme is being implemented through Panchayats and Nagarpalikas.

Public Distribution System (PDS) and the Targeted Public

Distribution System (TPDS)

A well targeted and properly functioning Public Distribution Systenl (PDS) is an important constituent of the strategy for poverty alleviation. PDS continues to be a major instrument of Government's economic policy for ensuring food security to the poor. With a network of more than 4.62 lakh fair price shops (FPS) distributing commodities worth more than Rs 30,000 crore annually to about 160 million families, the PDS in India is perhaps the largest distribution network of its kind in the world. For effective functioning of PDS, the Central Government bears the responsibility for procurement and supply of foodgrains, namely wheat and rice, besides sugar, imported edible oils and kerosene to the State governments and the Union Territories for distribution. Some States / UTs distribute additional items of mass consumption also through the PDS outlets.

The PDS as it was being implemented earlier, had been widely criticized for its failure to serve the population Belao the Poverty Line (BPL), its urban bias, limited coverage in the States with high concentration of the rural poor and lack of transparent and accountable arrangements for delivery. Therefore, in June 1997, the Government of India launched the Targeted Public Distribution System (TPDS) with focus on the poor. Under the new systern a two tier subsidized pricing system has been introduced to benefit the poor. The essential features of TPDS are: Government of India is committed to making available foodgrains to the States to meet the requirement of foodgrains at the scale of 10 Kg per month per family at specially subsidized prices to population falling below the officially estimated poverty line (BPL families). The states would also receive the quantity needed for transitory allocation to Above Poverty Line (APL) population. The state governments were to streamline the PDS by issuing special cards to BPL families and selling essential articles under TPDS to them at specially subsidized prices, with better monitoring of the delivery system.

The BPL households were determined on the basis of population projections of the Registrar General of India for 1995 and the State wise poverty estimates (1993-94) of the Planning Commission for 1993-94. The total number of BPL households so 'determined was 596.20 lakh. Thus the scheme, when introduced, was intended to benefit these poor families for whom a quantity of about 72 lakh tonnes of food grains was earmarked annually. Keeping in view the consensus on increasing the allocation of food grains to BPL families, and to better target the food subsidy, Government of India increased the allocation to BPL families from 10 Kg to 20 Kg of food grains per family per month at 50% of the economic cost and allocation to APL families at economic cost w.e.f. 1.4.2000. The allocation of APL families was retained at the same level as at the time of introduction of TPDS but the *Central Issue Prices* (CIP) for APL were fixed at 100% of economic cost from that date so that the entire consumer subsidy could be directed to the benefit of the BPL population.

The end retail police is fixed by the Statements after taking into account margins for wholesalers/retailers, transportations charges, levies, local taxes, etc. Under the TPDS the States are requested to issue food-grains at a difference of not more than 50 paise per Kg over and above the CIP for BPL families. Flexibility to States has been given in the matter of fixing the retail issue prices by removing the restriction of 50 paise per Kg over and above the CIP for distribution of food grains under TPDS except with respect to Antyodaya Anna Yojana (AAY) where the end retail price is to be retained at Rs. 2 /- per Kg for wheat and Rs. *31*- per Kg for rice. We shall learn about the AAY scheme in a little while from now.

For identification of BPL families under TPDS, Gram Panchayats and Nagar Palikas are involved. While doing so, the thrust is to include the really poor and vulnerable sections of the society such as landless agricultural labourers, marginal farmers, rural artisans craftsmen such as potters, tapers, weavers, black-smith, carpenters, etc. in the rural areas and slum dwellers and persons earning their livelihood on daily basis in the informal sector like potters, Rickshaw-pullers, cart-pullers, fruit and flower sellers on the pavement, etc. in urban areas.

Scale and Issue Price

Since 1997, the Scale of issue of the BPL families has been gradually increased from 10 Kg to 35 Kg per family-per ration month. The scale of issue was increased from 10 Kg to 20 Kg per family per family per month with effect from 1st April, 2000. The allocation for APL families has been retained as the same level as at the time of introduction of TPDS (i.e 10 Kg per family per month). The allocation of food grains for the BPL families has been further increased from 20 Kg to 25 Kg per family per month with effect from July, 2001. Initially, the Antyodaya (AAY) families were provided 25 Kg of food grains per family per month at the lime of launching of the scheme. The scale of issue under APL, BPL and *AAY* has been revised to 35 Kg per family per month with effect from 1st April, 2002 with a view to enhancing the food security at the household level.

Chapter 7

Energy Metabolism

Lavoisier is rightly considered as the pioneer in studies on energy metabolism. He along with Laplace, carried out, during the 18th century, experiments with guinea pigs and found a relationship between heat produced and carbon dioxide output. Later, Lavoisier also measured the oxygen consumed by men and found that exercise increased the oxygen consumption.

The human body utilizes the potential energy contained in foods for maintaining life and doing work. The energy yielding factors are Carbohydrates, Fats and Proteins. Within the body these nutrients are oxidized in the cells with the help of catalysts, such as enzymes, co-enzymes (containing certain vitamins and minerals) and hormones.

1. Carbohydrates (CHO)

Digestion

Digestion of carbohydrates begins in the mouth, where salivary amylase hydrolyzes starch into smaller polysaccharides and maltose. Only about 5% of the starch is hydrolyzed in the mouth. After the contents enter the stomach, acidity of the stomach stops the action of salivary amylase, by denaturing the enzyme. Thereafter, in the small intestine, the pancreatic amylase continues the digestion of starch. Various disaccharides (sucrose, maltase and lactase) break disaccharides into monosaccharides.

Dietary fiber moves to the large intestine, where bacteria partially ferment the undigested carbohydrate and produce gas and few short-chain fatty acids. Soluble fiber softens the tool and makes it easier to pass. Insoluble fiber resists bacterial activity and passes unchanged to add bulk to stool.

Absorption

Different monosaccharides are absorbed into the mucosal cell lining the small intestine, by two mechanisms. Whereas fructose is absorbed by facilitated diffusion, glucose and galactose are absorbed by active transport, by sodium linked glucose transporter (SLGT-1). Monosaccharides pass through the intestinal mucosal cells and enter the liver via the portal vein. Within the liver, galactose and fructose are also converted to glucose.

Storage

Liver stores glucose as glycogen. Glycogen is hydrolyzed to glucose, as and when body needs it. Excess of carbohydrate is also stored as fat.

Metabolism of Carbohydrates

The cells throughout the body use glucose to supply energy required to drive various chemical processes. Complete oxidation of glucose yields O_2, water and energy (ATP). This occurs in several steps. First, glucose is broken down to pyruvate by the process referred to as glycolysis. Thereafter, pyruvate looses one carbon and forms acetyl CoA, which enters citric acid cycle and electrons and hydrogen to electron transport chain to generate ATP. Glucose is also stored in the liver and muscle as glycogen.

Gluconeogenesis

When there is lack of dietary carbohydrate, liver and kidney can synthesize glucose from non-carbohydrate sources. This process is referred to as gluconeogenesis. Liver is the major site of gluconeogenesis and accounts for about 90% of glucose production. Kidneys also synthesize about 10% of glucose by this process. Major precursors of glucose are lactate, glycogenic amino acids and glycerol.

Glycogen synthesis

Body stores excess of glucose in liver and muscle as glycogen by the process called glycogenesis.

Glycogen breakdown

Glycogen is broken down to release glucose by the process called glycogenolysis.

2. Proteins

The capable proteins to performs their various functions, dietary protein needs to be broken down into its constituent amino acids. The mechanical digestion of protein begins in the mouth, where the teeth grind the food into small pieces. The mouth does not produce any enzyme to digest proteins. Chemical digestion begins in the stomach. The hydrochloric acid (HCL) in the gastric juice activates the enzyme pepsin, which acts on proteins and reduces them to polypeptides. After the partially digested proteins reach the small intestine, three pancreatic enzymes-trypsin, chymotrypsin, and carboxypeptidase-continue the process of chemical digestion. The peptidases secreted by the intestine finally reduce the smaller peptides and dipeptides into amino acids, which are the end-product of protein digestion.

After digestion, the amino acids in the small intestine are absorbed by the blood and carried to the liver and all body tissues, where they are metabolized. The body needs varying compositions of amino acids to build and repair the different tissues of the body. All essential amino acids must be present if body cells have to be build or repaired. Surplus amino acids are sent back to the liver, where they are dominated by splitting off the amino group. The remaining part of the protein is used for energy or stored as glycogen in the liver and muscles or as fat in the adipose tissue. The end products of the metabolism of amino acids are carbon dioxide, water, and nitrogen. Some of the nitrogen is excreted as urea by the kidneys. Some nitrogen may be retained and used again to synthesize non-essential amino acids, as and when required by the body.

3. Fat

The fats and oils consumed in the diet need to be broken down into their constituent components, namely glycerol and fatty acids, before they can be utilized by the body.

Digestion : No chemical digestion of fat takes place in the mouth. Fat is broken into smaller particles by chewing and is mixed with saliva. In the stomach, peristaltic movements churn the food along with the gastric juice. Two enzymes help in digesting fats, namely gastric lipase in the gastric juice and pancreatic lipase from the pancreas which acts in the small intestine. For fats to be digested, they first need to be emulsified. Gastric lipase acts on emulsified butterfat, and the main digestion of fats begins in the small intestine.

The presence of fats in the duodenum stimulates the secretion of bile from the gall bladder. Bile acts as an emulsifier and breaks down large fat globules into smaller particles. This increases the total surface area of fat and increases the efficiency of enzyme action. The alkaline nature of bile helps pancreatic lipase to remove fatty acids from the triglyceride, converting them to diglycerides and monoglycerides and finally to fatty acids and glycerol.

The final products of digestion to be absorbed are fatty acids, glycerol, monoglycerides, and diglycerides. Some remaining fat which is undigested may be excreted through the faeces.

Absorption: Since fats are insoluble in water, they cannot be directly absorbed into the intestines and blood stream without making them absorbable. Bile helps in absorption of fat by forming a complex with fatty acids and glycerides which is absorbed by the intestinal wall. Once absorbed, bile separates and returns to the intestine to recombine with fatty acids and glycerides and the process continues.

Metabolism: In the intestinal wall two important reactions take place:

The enzyme enteric lipase breaks down the remaining mono-and diglycerides into fatty acids and glycerol.

The fatty acids and glycerol recombine to form new body fats or newly formed triglycerides which need to be absorbed and transported via the blood circulation. The newly formed triglycerides and fats are covered with small amounts of protein to form lipoproteins called chylomicrons. Chylomicrons enter the lymphatic circulation from lymph vessel in the small intestine. From the lymphatic system, they enter into the portal circulation and are converted into other lipoproteins in the liver. These lipoproteins are circulated to all body cells for energy and other structural functions, or stored in the adipose tissue as a reserve source of energy.

4. Energy

Energy value of food

The energy content of various foods can be measured in two ways: by calorimetry or by proximate composition.

Calorimetry

The bomb calorimeter is based on the principle of calorimetry. It measures the heat produced when the food sample is ignited by an electric spark in the presence of oxygen and platinum which acts as a catalyst. The calorimeter is made up of two main parts - an inner part in which a measured quantity of food sample to be tested is placed and an outer portion which contains a known volume of water. When the food sample is electrically ignited, the surrounding water absorbs the heat produced. The energy value of the food is calculated by measuring the rise in temperature of water, based on the definition of the term calorie.

Proximate composition

This is a simple and faster way of determining the calorie content of food. The approximate carbohydrate, fat, and protein content of a food given in the food composition tables is multiplied by their fuel factor, *i.e,* calories provided by 1 g of the nutrient and this total gives the energy value of a food .The energy content of food given in the food composition tables are derived values. The carbohydrate content is calculated by difference, subtracting water, minerals, proteins, fats, and crude fiber content.

The potential energy present in food when measured in a bomb calorimeter is higher than the energy released in the body. This is because some losses occur during digestion, absorption, and metabolism. The potential energy value must be corrected to allow for losses that occur during digestion, absorption, and metabolism of nutrients. The physiological fuel factors are based upon the corrections for losses of unabsorbed nutrients in the faeces and nitrogenous products excreted in the urine. On a mixed diet containing high-quality proteins, 85-90 per cent carbohydrate, 95 per cent carbohydrate, 95 per cent fat, and 92 per cent proteins are digested and absorbed.

The carbohydrates and proteins which are absorbed are fully oxidized. However, when proteins are oxidized as a source of energy, they first need to be deaminated (removal of amino group). The amino group is converted to urea and is excreted in the urine. Hence, the body loses a part of the protein that is digested and absorbed. This loss of protein is about 25 per cent. The net energy value obtained from food by the body is lesser than that measured in bomb calorimeter.

Energy needs of the body

The energy requirement for an average healthy person is based on the energy required to carry out basal processes as well as the energy cost of activities one indulges in. The average requirements for both the sexes belonging to different age categories is given on recommended dietary allowances (RDA) .These allowances are suggested for a reference man weighing 60 kg and a reference woman weighing 50 kg in the 20-39 year age group.

Reference man: The reference man is between 20-39 years of age and weighs 60 kg. He is in a state of good health, free from disease, and physically fit for active work. He is moderately active and is employed for 8 hours in moderate work. He spends 8 hours in bed, 4-6 hours sitting or engaged in light work, and 2 hours in walking, active recreation, or household chores. On an average, his energy requirement is 2875 kcal.

Reference woman : The reference woman is between 20-39 years of age, healthy, and weighs 50 kg. She is moderately active and spends 8 hours in routine household work or in light work. She spends 8 hours in moderate work. she spends 8 hours in bed, 4-6 hours sitting or moving around and performing light activity, and 2 hours in walking or active recreation or household chores. On an average her energy requirement is 2,225 kcal.

Carbohydrates, fats, and proteins provide us with energy. About 60-65 per cent of the total calorie requirement should be from carbohydrates. The energy from fat should not exceed 30 per cent, preferably 15 per cent calories from fat of which 15 g should be polyunsaturated vegetable oils to meet the requirement for essential fatty acids; and 10-15 per cent calories from proteins.

Total energy requirement

The total energy required by a person is the sum total of basal energy needs, the energy effect of food intake or the specific dynamic action and energy cost of physical activity. Energy is needed for growth, for maintenance, for the innumerable processes continuously taking place, for regulating body temperature, and for physical and mental activity. Activities that need energy are broadly classified into the following:

1. **Voluntary activities:** e.g., activities under the control of our will such as walking, sitting, cycling and dish washing.
2. **Involuntary activities**: that go on irrespective of whether we want them to. They are not under the control of our will and are vital activities on which our survival depends, such as beating of the heart, respiration, and maintaining body temperature. Energy is first provided for these activities and is referred to as basal metabolism.

   Total energy = Basal metabolic + Specific rate + Activity

   Needed by the body dynamic action.

Basal metabolic rate

The amount of energy required by the body for carrying out involuntary work and maintaining the body temperature is known as the basal metabolic rate (BMR). The involuntary work includes the functioning of various organs and systems which work continuously to keep the body processes going, such as the heart and blood circulation, the kidneys and excretion. Approximately one-third of this energy is needed for these processes, while the remaining two-thirds is utilized for maintenance of muscle tone.

Test for Basal Metabolism

To measure the BMR, the following conditions need to be observed.

1. The test is conducted before breakfast 12-16 hours after the last meal, i.e., in a post-absorptive state, to eliminate the influence of food.
2. The subject should be relaxed and lying down but awake because sleep reduces BMR and activity or exercise increase the rate of oxidation the cells.

3. The subject should have a normal pulse rate and be free from tension or fear of the test.
4. The ambient temperature should be comfortable as perspiration or shivering may effect the rate of oxidation.
5. The subject should be afebrile as fever increases the BMR by 7 per cent for every degree Fahrenheit rise in temperature.

The test is conducted by measuring the amount of oxygen consumed and/or carbon dioxide exhaled in a predetermined time, usually-6-8 minutes.

A deviation of 10-15 per cent from the accepted standards may be considered normal. The adult BMR is 1 kcal/kg body weight/hour for men and 0.9 k cal/kg body weight/hour for women, or approximately 1,440 kcal for men and 1,080 kcal for women, respectively.

Factors affecting the BMR: Body size and composition, age and growth, sleep, state of nutrition, climate and certain ailments affect the basal requirement.

1. **Body size and composition:** The heat loss from the body is related to body size; energy needed to maintain lean muscle mass at rest is related to body composition.

   Athletes, who have developed muscles due to exercise, have 5 per cent increase in BMR as compared to non-athletes. Women have 5 to 10 per cent lower metabolic rate as compared to men of same weight and height, because they have more fat and less muscle in their body than men do. If BMR is calculated on the basis of lean body mass, there is no difference between men and women.

2. **Age.**: There is a decrease in metabolic energy expenditure of 2-3 per cent per decade after early adulthood, due to the shift in the proportion of muscle to fat in the body. The basal metabolic rate gradually decreases after reaching adulthood; the decrease is about 30 per cent between 30 and 75 years of age.

3. **Growth and repair:** The metabolic rate is highest in the stages of rapid growth, namely the first and second years of life. The metabolic rate increases in pregnancy due to growth of the foetus and related increased growth activity.

4. **Sleep:** The metabolic rate falls by about 10 per cent while sleeping. This is due to relaxation of muscles and reduced activity of the nervous system during sleep.

5. **State of Health**: The metabolism is decreased due to malnutrition; the decrease in basal metabolism is proportional to the degree of malnutrition. It is mainly due to decrease in the amount of active tissue and a decrease in the metabolic rate. Fever, the increase in temperature of the body, increases metabolism. The increase is about 7per cent per each degree Fahrenheit above normal (98.6°F or 37°C).

6. **Hormonal control** : A hormone known as thyroxine, controls the speed of our involuntary activity. The thyroid gland, situated in our neck, synthesizes this hormone. If too much thyroxine is released, the rate of energy expenditure is increased; if too little is released, the energy expenditure is reduces. The basal metabolism may decrease by 30 to 40 per cent, when the synthesis of thyroxine is inadequate. On the other hand, the BMR may almost double due to a hyperactive thyroid gland.

7. **Climate** : The BMR rises when the climate is cold in order to maintain normal body temperature. In very warm climates leading to profuse sweating, BMR may increase by trying to reduce body temperature.

8. **Physiological tension** : Worry and anxiety increase BMR.

Energy for physical activity

The total energy needed for physical activities includes that spent in personal necessities, professional work and leisure activities. The additional energy output due to voluntary physical activity varies from as little as 10 per cent for a bed-ridden patient to as high as 50 per cent for an athlete. Factors, which affect energy need for activities, are body size, the efficiency in activity and level of fitness of the person.

Table 21: Recommended daily allowances for energy

Particulars	K.cal/day	Particulars	Kcal/day
Man (60 kg)		**Infants**	
Sedentary work	2320	0+6 months (5.4 kg)	90 Kcal/kg
Moderate work	2730	0+6 months (5.4 kg)	of body wt.
Heavy work	3490		80 Kcal/kg
			of body wt.
Woman (55kg)		**Children**	
Sedentary work	1900	1-3 years (12.2 kg)	1060
Moderate work	2230	4-6 years (19.0 kg)	1350
Heavy work	2850	7-9 years (26.9 kg)	1690
Pregnant woman	+350	Boys 10-12 years (35.4 kg)	2190
Lactating woman		Girls 10-12 years (31.5 kg)	2010
0.6- months	+600	Boys 13-15 years (47.8 kg)	2750
6-12 months	+520	Girls 13-15 years (46.7 kg)	2330
		Boys 16-18 years (57.1 kg)	3020
		Girls 16-18 years (49.9 kg)	2440

Source : Nutritive value of Indian foods, 2010, ICMR, NIN, Hyderabad

1. Energy for utilization of food : Some energy is used to take food into the body, digest it and carry the nutrients to the tissues and eliminate the waste products produced in the process. This is known as the specific dynamic action of foods or the energy for utilization of food.

2. Energy for growth and repair: Extra energy is needed during periods of rapid growth for the building of new tissues. These stages of life include pregnancy, early childhood and adolescence.

Energy balance

The human body is constantly using energy which needs to be replaced. For this, a constant supply of energy is required. Energy is used for basal metabolism, specific dynamic action, and physical activity. The energy from the food we eat, mainly carbohydrates and fats, is used to meet the energy demands of the body. When food is not available during fasting or starvation, the body has three types of energy store.

Glycogen: The form in which carbohydrate is stored in the muscles and liver is adequate to last for 12-48 hours. Approximately 300 g glycogen is stored in the muscle and 100 g in the liver.

Muscle Protein is stored in limited amounts in the muscle.

Adipose tissue : Fat is stored in the adipose tissue and the amount stored varies vastly from one person to another.

A person in energy balance neither gains weight nor loses weight. Excessive consumption of calories as compared to the output or activity leads to a condition called overweight, which in severe cases is called obesity. A deficient intake of carbohydrates and fats in the diet leads to underweight or under nutrition. Both underweight and obesity are undesirable conditions which need timely correction.

Definition : Energy balance is a condition in which the energy provided by food is nearly equal to the total energy expended by the body resulting in a steady body weight.

Energy balance: Energy output = Energy input

Calorific value of Carbohydrates, Fats and Proteins

The average Calorific values of pure carbohydrates, fats and proteins determined with the bomb calorimeter are given below:

1 g Carbohydrates – – – 4 Kcal

1 g Fats – – – – – – – – – – – – 9.45 Kcal

1 g Protein – – – – – – – – – – 5.65 Kcal

Physiological energy value of foods

In the Bomb calorimeter, Carbohydrates and Fats are completely oxidized to CO_2 and water. Protein is oxidized to CO_2, Water and Nitrogen. Another important error in the use of Bombcalorimeter, for determining the calorific value of foods of vegetable origin is that the fiber present in foods is burnt and yields energy, while it is not utilized by human beings. But, in the utilization of carbohydrates, fats and proteins in the body, a certain percentage is lost in digestion and the nitrogen in proteins is excreted mainly as urea, which contains some energy value.

Loss in Digestion : The average losses of carbohydrates, fats and proteins during digestion in human beings on diets consumed in western countries have been estimated to be as 2 per cent for carbohydrates, 5 per cent for fats and 8 Percent for proteins. Food energy available after loss in digestion is shown in Table 22.

Loss of energy in metabolism due to incomplete oxidation: There is no loss in metabolism in the case of carbohydrates and fats .But in the case of proteins, a part of the energy is lost as urea due to incomplete oxidation. This loss has been estimated as 1.2cal/g of protein oxidized. The physiological energy values of carbohydrates, fats and proteins are 4, 9 and 4 after making allowances for lossess of food energy in digestion amd metabolism (Table 22). These values are known as Atwater-Bryant factors and they are used widely in all countries for calculating energy values of foodstuffs from their chemical composition.

Table 22: Calculation of Physiological energy value of carbohydrates, fats and proteins from Gross energy value

Nutrients	Gross energy value (K cal/g)	Loss of food energy in diges-tion %	Energy available after digestion (K cal/g)	Loss of food energy in metabolism (K cal)	Physiological energy value of foods (K cal/g)
Carbo-hydrates	4-1	2	4-0	Nil	4-0
Fats	9-45	5	9-0	Nil	9-0
Proteins	5-65	8	5-2	1-2	5-0

Calculating the energy value based on proximate principles

Problem: A non-vegetarian meal provides 20 g protein, 20 g fat, and 125 g carbohydrate, of which 5 g is fiber. Calculate the calories provided by this meal.

1. Protein

1 g protein provides 4 kcal

Therefore, 20 g protein will provide 20x 4 = 80 kcal.

2. Fat

1 g fat provides 9 kcal

Therefore, 20 g fat will provide 20 x 9 = 180 kcal

3. Carbohydrate

1 g carbohydrate provides 4 kcal

1 g fiber provides 0 kcal

Digestible carbohydrate = 125-5=120 g

120 g carbohydrate will provide 120 x 4 = 480 kcal

Total calories provided by the meal = 80 + 180 + 480
= 740 kcal

The meal provides 740 kcal.

Sources of energy in the diet

	Kcal/g	kJ/g
Protein	4	17
Carbohydrate	4	17
Fats	9	38
Alcohol	7	29

Dietary Sources

All foods provide energy. While selecting food, one must consider the other nutrients, such as proteins, vitamins, minerals, and fiber present in the food and make a wise choice. The nutrients carbohydrates and fats are consumed mainly as a source of energy. All refined carbohydrates, i.e., sugars and starch and all foods rich in fats are rich sources of energy. The cereal group is another excellent source of energy and supplies the highest percentage of calories. Hollow-calorie food or empty calories are provide by food which is rich in energy but lack other vital nutrients. Sugars, fats, and alcoholic beverages are referred to as empty calorie food because they contain traces of vitamins and minerals. They are low in nutrient density. Nutrients-dense food are those food which are rich in one or more nutrients apart from calories. It is a measure of the quantity of these nutrients supplied by a food in relation to its calorie content. Nutrients-dense foods are of special importance to weight watchers and those leading a sedentary lifestyle. It helps them get all the nutrients they need without consuming unnecessary calorie-rich foods and gaining weight.

Estimation of energy requirements

The total energy required in a day can be estimated by two methods: RDA tables and actual record of activities.

The RDA table gives us the approximate calories required on the basis of age, sex, and activity. A specific estimate of one's energy needs can be calculated by adding energy cost of various activities is mentioned. If the time taken for each activity is known, the energy cost can be easily calculated. A sum total of energy required for BMR +SDA+Physical activity could also give us the total energy requirement.

Calculating Energy Requirements

Particulars - Adult woman, 25 years old, weight 50 kg, Height 155 cm

- Estimate your level of physical activity for a heavy worker will be :

Activity	Energy cost as a percentage of BMR
Sedentary	20 percent
Moderate	35 per cent
Heavy	50 percent

Energy cost of physical activity for a heavy worker will be

= (50 per cent x BMR)+ BMR

= 50/100 x 1,080 + 1,080

= 1,620 kcal

Add energy cost of food intake (SDA) which is 6-10 per cent of total calories. On a vegetarian diet it is about 6 per cent and on a non-vegetarian diet it is about 10 per cent, approximately 100 k cal.

Total energy requirement = BMR + Physical activity + SDA = 1,080 + 1,0620 + 100 = 2,800 kcal

Modifying energy content of meals

The energy content of the diet can be suitably modified to enable a person to gain or lose weight. It is essential to understand the causes of weight loss or weight gain before suggesting modifications. The weight gain or weight loss should be gradual.

Under weight

The cause of weight loss must be treated before a modified diet is prescribed.

To gain weight

- High-calorie diet, i.e 500 kcal more per day for weight gain of 1 lb /week
- High-protein diet with good quality proteins 1.2 g/kcal desirable body weight
- Adequate vitamins and minerals. Sometimes supplements may be necessary
- Modify diet gradually
- Have six small meals a day
- Keep meal times pleasant and food should be appetizing and attractively served.

Type of foods recommended

- Food which is easy to assimilate
- High calorie, high-proteins foods
- Salad dressing, cream soups
- Desserts
- Fresh fruits such a banana, mango, chikoo
- Whole milk, whole milk curd and paneer
- Nutritious soups and stews
- Animal protein-baked fish, mutton stew, cheese sandwiches

Overweight/obesity

To reduce weight

- Low-calorie, high-fiber, high -proteins diet
- Low fat, no salad dressings
- Low sugar and refined carbohydrate
- Avoid sugar in tea and coffee
- Use lean meat
- Have three meals a day

Types of foods recommended : Foods which give a feeling of fullness at the end of a meal, i.e. low energy density meals, lots of substance but lesser calories.

To create satiety, increase

1. Bulk or fiber content
2. Volume

Fats, butter, and oil should be used sparingly by an obese individual.

Foods suggested

- Soups: Large portion of liquid, rich in ingredients. The steamy aroma gives a sense of satisfaction.
- Vegetable juices: Served before a meal reduces quantity of food consumed at mealtime.
- Salads: A salad platter looks like a large portion of food served. It takes time to chew, has lots of fiber, and is low calorie. It is part of a low-energy-density meal recommended for all individuals. It is specially included in a weight reduction diet.
- High-fiber breakfast cereal: It requires time to eat and chew keeps blood sugar levels steady.
- Plain lassi made from skimmed milk and served diluted and unsweetened could be had before or between a meal.
- Smoothies Appetite suppressing fruit protein drinks. The volume is increased by whipping fruit and yoghurt/soya curd/protein powder in the blender with crushed ice for a longer time.

❑❑❑

Chapter 8

Balanced Diet and Modified Diet

Balanced Diet

Nutrients that we obtain through food have vital effects on physical growth and development, maintenance of normal body function, physical activity and health. Nutritious food is, thus, needed to sustain life and activity. Our diet must provide all essential nutrients in the required amounts. Requirements of essential nutrients vary with age, gender, physiological status and physical activity. Dietary intakes lower or higher than the body requirements can lead to under-nutrition (deficiency diseases) or over-nutrition (diseases of affluence) respectively. Eating too little food during the vulnerable periods of life such as infancy, childhood, adolescence, pregnancy and lactation and eating too much at any age can lead to harmful consequences. An adequate diet, providing all nutrients, is needed throughout our lives. The nutrients must be obtained through a judicious choice and combination of a variety of foodstuffs from different food groups (Fig. 11).

Fig. 11: Food Pyramid

→ Foods high in fats and sugars: Eat sparingly
→ Meat, fish and dairy: Eat moderately
→ Fruit and vegetables: Eat liberally
→ Carbohydrates (cereals millets, pasta, bread, potatoes): Eat adequately

Balanced diet

A balanced diet is one which provides all the nutrients in required amounts and proper proportions. It can easily be achieved through a blend of the four basic food groups. The quantities of foods needed to meet the nutrient requirements vary with age, gender, physiological status and physical activity. A balanced diet should provide around 50-60% of total calories from carbohydrates, preferably from complex carbohydrates, about 10-15% from proteins and 20-30% from both visible and invisible fat. In addition, a balanced diet should also provide other non-nutrients, such as dietary fiber, antioxidants and phytochemicals which bestow positive health benefits. Antioxidants such as vitamins C and E, beta-carotene, riboflavin and selenium protect the human body from free radical damage. Other phytochemicals such as polyphenols, flavones, *etc.*, also afford protection against oxidant damage. Spices like turmeric, ginger, garlic, cumin and cloves are rich in antioxidants.

Difference between an adequate diet and a balanced diet

- An "**adequate diet**" includes sufficient energy for the person's needs - though the energy in the diet may be in any form, e.g. as carbohydrate, protein, fat.
- A "**balanced diet**" not only includes sufficient energy for the person's needs but all of the person's dietary requirements in the correct proportions.

Food Groups

Foods are conventionally grouped as :

1. Cereals, millets
2. Pulses and legumes
3. Vegetables and fruits
4. Milk and milk products, egg, meat and fish
5. Oils & fats and sugars

However, foods may also be classified according to their functions (**Table 23**)

Components of a balanced diet: A balanced diet must provide the building blocks for growth and repair, and sufficient energy. There are seven components to a balanced diet:

Carbohydrates - The majority of energy intake should come from these molecules, which provide an energy source – these can be divided into sugars and starches.

Proteins - It is needed for growth and repair, and are a minor energy source (55% of energy should come from carbohydrates, 35% from fats and only 10% from protein), due to it being a less efficient energy source.

Fats - major energy reserve/resource, used for insulation and cell membranes, hormones and cholesterol.

Fiber - technically a carbohydrate but we don't have the enzyme to break it down, so it is indigestible, but is an essential part of the diet for ensuring a functioning digestive system - fiber aids the passage of food through the alimentary canal (also the breakdown by bacteria of fibers can help provide certain trace vitamins).

Water - An important transport medium, water is essential for the body to function, and it is involved directly in some metabolic reactions.

Vitamins - These play an important roles in the chemical processes taking place inside cells - some vitamins are water-soluble and others are fat-soluble.

Minerals - These are the inorganic elements occurring in the body that are essential to its normal functions.

Table 23: The five food groups and their major nutrients

Food group	Main nutrients
1. **Cereal grains and products**	
Rice, Wheat, Ragi, Bajra, Maize, Jowar, Barley, Rice flakes, wheat flour.	Energy, Protein, Invisible fat, Vitamin-B1, Vitamin-B2, Folic acid, Iron, Fiber
2. **Pulses and Legumes**	
Bengalgram, Blackgram, Greengram, Redgram, Lentil (whole as well as dhals), Cowpea, Peas, Rajmah,	Energy, Protein, Invisible fat, Vitamin-B1, Vitamin-B2, Folic acid, Calcium, Iron, Fiber.
Soyabean, Beans.	Protein, Calcium,Iron
3. **Milk and Meat Products**	
Milk, Curd, Skimmed Milk, Cheese, Chicken, Liver, Fish, Egg, Meat.	Protein, Fat, Vitamin-B2, Calcium
4. **Fruits and Vegetables**	
Fruits: Mango, Guava, Tomato, Papaya, Orange, Sweet lime, Water melon.	Carotenoids, Vitamin-C, Fiber, Invisible fat, Vitamin-B2, Folic acid, Iron.
Green leafy Vegetables:	Carotenoids, Vitamin-B2, Folic acid,

Amaranth, Spinach, Gogu, Drumstick leaves, Coriander leaves, Fenugreek leaves.	Calcium
Other Vegetables: Carrots, Brinjal, Lady's finger, Beans Capsicum, Onion, Drumstick, Cauliflower.	
5. **Fats and Sugar**	
Fats: Butter, Ghee, Hydrogerated fat, Cooking oils like groundnut, Mustard, Coconut	Energy, Fat, Essential fatty acids.
Sugar: Jaggery and sugar	Energy

Table 24: Classification of foods based on function

Classification of foods	Food sources	Nutrients
Energy rich foods	**Carbohydrates & fats** Whole grain cereals, millets Vegetable oils, ghee, butter Nuts and oilseeds Sugars	 Protein, fiber, minerals, calcium, iron &B-complex vitamins Fat soluble vitamins, essential fatty acids
Body building foods	**Proteins** Pulses, nuts and oil seeds Milk and Milk products Meat, fish, poultry	 B-complex vitamins, invisible fat, fiber Calcium, Vitamin A, Riboflavin, vitamin B
Protective foods	**Vitamins and Minerals** Green leafy vegetables Other vegetables and fruits Eggs, milk and milk products and flesh foods	Antioxidants, fiber and other Carotenoids Fiber, sugar and antioxidants Protein and fat

Table 25: Balanced diet for men and women (NIN)

Food group (g)	Adult men			Adult Women		
	Sedentary work	Moderate work	Heavy work	Sedentary work	Moderate work	Heavy work
Cereals	460	520	670	410	440	675
Pulses	40	50	60	40	45	50
Leafy vegetables	50	40	40	100	100	50
Other vegetables	60	70	80	40	40	100
Roots and tubers	50	60	80	50	50	60
Milk(ml)	150	200	250	100	150	200
Oils and fats	40	45	65	20	25	40
Sugar and Jaggery	30	35	55	20	20	40

Additional diet is required during pregnancy and lactation

Pregnancy is a demanding physiological state. In India, it is observed that diets of women from the low socioeconomic groups are essentially similar during pre-pregnant, pregnant and lactating periods. Consequently, there is widespread maternal malnutrition leading to high prevalence of low birth weight infants and very high maternal mortality. Additional foods are required to improve pregnancy weight gain and birth weight of infants, Pre-pregnant BMI, maternal age and rate of pregnancy weight gain must be considered in tailoring the calorie recommendation to the pregnant women.

A balanced diet should provide around 60–70% of total calories from carbohydrate, 10-12% from protein and 20–25% of total calories from fat.

Balanced diet

- Meets nutritional requirement
- Prevents degenerative diseases
- Improves longevity
- Prolongs productive life

- Improves immunity
- Increases endurance level
- Develops optimum cognitive ability
- Helps in coping up stress.

Thus balanced diet enhances quality of life. Balance diet should meet the nutritional requirements as per the age, body composition, health condition and BMI also plays important role in planning balance diet.

My plate

Fig. 12: My Plate

The U.S. Department of Agriculture (USDA) created *My Plate,* an easy-to-follow food guide, to help parents to figure out how to feed their children. The coloruful divided plate includes sections for vegetables, fruits, grains, and foods high in protein. It's an improvement upon the complex and confusing My Pyramid that the USDA once used to explain its dietary guidelines (Fig. 12).

With My Plate, forget about measuring out serving sizes and trying to remember whether a piece a chicken should be the size of a fist or a golf ball. My Plate's user-friendly, interactive website provides simple messages that are easy to understand, such as:

- Fill half plate with vegetables and fruits

- Make at least half the grains we serve whole grains,
- Serve fat-free or low-fat (1%) milk and water rather than sugary drinks
- When buying pre-packaged foods, choose ones that are low in sodium
- Don't serve oversized portions.

Changes from Pyramid to Plate

My Pyramid represented the five food groups plus oils with six colourful vertical stripes. The stripes were thicker or thinner depending on how much of one food group should be consumed. Steps on the left side of the pyramid reminded people to stay active each day.

At first glance, it was difficult to tell how much of each food group should be in a healthy diet. That's where My Plate really makes meal planning easier. Just by looking at the icon, we know right away that vegetables and fruits should take up half the plate, and grains and protein foods should take up the other half, And with a side helping of dairy, we reminded to include milk or another dairy food (like cheese or yogurt) in your daily meal plan.

Because My Plate is a divided plate, no one food group overpowers the others. That's because dietary guidelines encourage eating a variety of foods and discourage "super-sized" portions, which can lead to weight gain and obesity.

Oils provide important nutrients and are recommended in small amounts but aren't included on the My Plate icon. Choose oils over solid fats, but limit the amount you eat.

2. Modified diets

Purpose of diet therapy

- Maintain normal nutrition and health, e.g., modified diets are based on balanced diets.
- Treat deficiency diseases, e.g., high-protein, diet in protein calorie malnutrition.
- Alter nutrient requirement according to body's ability to use the nutrient, e.g., modify carbohydrates in a diabetic diet.

- Give rest to an organ or to the body e.g., intravenous fluids in severe vomiting and diarrhoea.
- Change body weight, e.g., low-calorie diet in obesity to lose weight.

The clinical dietitian in a hospital is responsible for planning therapeutic diets of patients as per the doctors diet prescription. The administrative dietitian manages food served establishments.

Dietetics: It is the science and art of feeding individuals or groups under different health and economic conditions according to the principles of nutrition and management.

Classification of modified diets

The diet is normally altered or modified in the following ways:

1. Modification in consistency
2. Modification in nutrient content
3. Modification in quantity
4. Modification in method of feeding

Nutrients	Disease	Causes	Symptoms
Proteins	Kwashiorkor	Deprived of breast milk	- Stunted growth - Succumb to infection - Loss of weight, anaemia - Persistence of infection
	Marasmus	Continued deficiency of calories and protein	- Shrivelled skin, monkey face - Dull and dry hair - Dehydration, diarrhoea - Low pulse rate - Subnormal temperature - Continued feeling of hunger
Carbohydrates	—	—	Carbohydrate deficiency in the diet is not likely to occur except in times of food shortage
Fats	—	Prolonged deficiency of fat in the diet	- Dryness of the skin - Secondary deficiency of fat soluble vitamins A, D, E, K.
		Excess fat in the diet	- Obesity - Digestive disturbances specially in children -Animal fats contain cholesterol

Minerals			
Calcium	Rickets in children, Osteomalacia, decay of teeth, excessive bleeding due to delayed coagulation	Deficiency of calcium	- Knock knee - Bowed legs - Stunted growth, soft bones -Enlargement of the ankles and wrists - Curved spine and pelvis
Iron	Anaemia	Prolonged deficiency	Prevents the formation of haemoglobin
Sodium	Heart cramps	Loss of sodium chloride by excessive sweating	- Intense and painful contractions of skeletal muscles
	Hypertension, toxemia	Due to excessive sodium during pregnancy	- Mental tension - Weak heart, vomiting etc.
Phosphorus	The effect of deficiency phosphorus is very rare in human beings because diets having sufficient cereals are seldom inadequate in this nutrient		
Iodine	Goitre	People living in hilly regions	Enlargement of the thyroid gland
Fluorine	Dental caries, tooth decay, dental fluorosis	Deficiency in growing period	Damage to the teeth and bones, loss of appetite, pain in pelvis, spine and limbs

Modified diets

Consistency

- Clear liquid
- Full liquid
- Soft
- Mechanical soft
- Light
- Pureed
- Bland

Nutrient Content

- High protein
- Low Protein
- Sodium restricted
- Low fat
- High fibre

Quantity

- Obesity
- Vomiting
- Diarrhoea
- Diabetes

Special Method of Feeding

External (by way of entry on or intestine)
Tube feeding e.g., formula diet

Parentaral
(by other routes not via the intestine)
IV fluids e.g., saline 5% dextrose

Fig. 13: Classification of modified diet

Menu planning

Definition

Menu planning is defined as a simple process which involves application of the knowledge of food, nutrients, food habits, and likes and dislikes to plan wholesome and attractive meals.

The caterer who is responsible for providing meals has to decide on various aspects, such as :

1. Menu
2. Serving size
3. Food cost
4. Suppliers and quantities to be purchased
5. Standardized recipes to be followed
6. Type of service Meal timings
7. Clientele

The aim of menu planning are to (Fig. 12a):

1. Meet the nutritional needs of the individuals who will be consuming the food
2. Plan meals within the food cost
3. Simplify purchase preparation, and storage of meals
4. Provide attractive, appetizing meals with no monotony
5. Save time and money.
6. Minimize overhead expenditure, *i.e.*, fuel, electricity, water, labour.

Menu planning is the most important aspect of planning and organization in food industry. It is an advance plan of a dietary pattern over a given period of time.

Factors influencing meal planning

Many factors influence the acceptability of a meal. Customers select what appeals most to them from a menu card based on individual likes and dislikes, budget, popularity of items, etc. However, while planning meals the following factor need to be considered.

Nutritional Adequacy

The most important consideration in menu planning is to ensure that the meal fulfills the nutrient needs of the individual consuming the meal. For example, if the meal is planned for an industrial worker, it must meet the recommended dietary allowances (RDA) for that age group. Foods from all basic food groups should be included in each meal so that the meal is balance and nutritional adequate.

Economic considerations: The spending power of the clientele has to be kept in mind and meals have to be planned within the budget.

Type of food service : Menus should be planned in relation to the type of food service, whether it is cafeteria, seated service, buffet, etc.

Equipment and work space : The menu should be planned keeping the available equipment and work space in mind.

Left over Food : An effective manager should consider as to how leftovers could be rotated to obtain maximum profit. Adequate storage space and hygienic standards should be ensure to minimize the risk of contamination and spoilage of food.

Food habits: Food habits of the customer is another important criteria which needs to considered as food served has to be acceptable to the customer.

Availability: Some fruits and vegetables are seasonal.

Seasonal fruits and vegetables should be given preference. Regional availability influences menu planning. For example, fish and sea food is fresh and cheaper in coastal areas.

Meal frequency and pattern: The meal timings and number of meals consumed in a day, whether meals are packed or served at the table, also influences the selection food items on the menu. The ate, activity level, physiological state, work schedule, and economic factors need to be known before planning meals for institutional catering.

Variety : This is one of the most important consideration while planning meals. A variety of foods from the different food groups should be included.

Planning balanced Meals

Meal planning involves proper selection of food to ensure balanced meals.

Food groups help us in planning balanced meals which supply all essential nutrients.

The three main functions performed by food are:

1. Providing energy
2. Body building and maintenance
3. Regulation of body processes and protection against infection.

On the basis of functions performed, food is classified into the following three groups:

1. Protective/Regulatory Foods
2. Body-building foods
3. Energy-giving foods

Steps in planning balanced meals

1. Collect information regarding the customer with respect to:
   - Age
   - Gender
   - Activity level
   - Religion
   - Socio-economic background
   - Food habits
2. Check the RDAs for energy and proteins
3. Prepare a food plan, i.e., list number of servings from each food group to meet the RDA
4. Decide on number of meals
5. Distribute servings for each meal
6. Select foods within each group and state their amount
7. Plan a menu
8. Cross-check to ensure that all food groups are included in requisite amounts. Using the above steps, plan a balanced diet for a day.

Table 26: Recommended Dietary Allowances for Indian as per 2010 guidelines by NIN, Hyderabad

Group	Particulars	Body Wt. (kg)	Net Energy (kcal/d)	Protein (g/d)	Visible fat (g/d)	Calcium (mg/d)	Iron (mg/d)	Vitamin A (µg/d)		Thiamine (mg/d)	Riboflavin (mg/d)	Niacin – equivalent (mg/d)	Pyridoxine (mg/d)	Ascorbic acid (mg/d)	Dietary folate (µg/d)	Vit. B12 (µg/d)	Magnesium (mg/d)	Zinc (mg/d)
								Retinol	β Carotene									
Man	sedentary work	60	2320	60	25	600	17	600	4800	1.2	1.4	16	2	40	200	1	340	12
	moderate work		2730		30					1.4	1.6	18						
	heavy work		3490		40					1.7	2.1	21						
Woman	sedentary work	55	1900	55	20	600	21	600	4800	1	1.1	12	2	40	200	1	310	10
	moderate work		2230		25					1.1	1.3	14						
	heavy work		2850		30					1.4	1.7	16						
	pregnant women		350	82.2	30	1200	35	800	6400	0.2	0.3	2	2.5	60	500	1.2		12
	lactation 0-6m		600	77.9	30	1200	25	950	7600	0.3	0.4	4	2.5	80	300	1.5		
	lactation 6-12 m		520	70.2	30					0.2	0.3	3	2.5					
Infants	0-6 months	5.4	92 kcal /kg /d	1.16 g /kg /d		500	46 µg /kg /d	350	2800	0.2	0.3	710 µg /kg	0.1	25	25	0.2	30	
	6-12 months	8.4	80 kcal /kg /d	1.69 g/ kg /d	19		5			0.3	0.4	650 µg /kg	0.4				45	
Children	1-3 years	12.9	1060	16.7	27	600	9	400	3200	0.5	0.6	8	0.9	40	80	0.2-1.0	50	5
	4-6 years	18	1350	20.1	25		13			0.7	0.8	11	0.9		100		70	7
	7-9 years	25.1	1690	29.5	30		16	600	4800	0.8	1	13	1.6		120		100	8
Boys	10-12 years	34.3	2190	39.9	35	800	21	600	4800	1.1	1.3	15	1.6	40	140	0.2-1.0	120	9
Girls	10-12 years	35	2010	40.4	35		27			1	1.2	13	1.6				160	9
Boys	13-15 years	47.6	2750	54.3	45		32			1.4	1.6	16	2	40	150	0.2-1.0	165	11
Girls	13-15 years	46.6	2330	51.9	40		27			1.2	1.4	14	2				210	11
Boys	16-17 years	55.4	3020	61.5	50		28			1.5	1.8	17	2	40	200	0.2-1.0	195	12
Girls	16-17 years	52.1	2440	55.5	35		26			1	1.2	14	2				235	1

Source: Anonymous, 2010. Nutrient requirements and recommended dietary allowances for Indians, National Institute of Nutrition, ICMR, Hyderabad.

Chapter 9

New Trends in Food Science and Nutrition

The development of food, nutrition and processing technology has been influenced by numerous factors; among them, consumer demands have undoubtedly oriented the new trends in the manufacturing, preservation, and control of food as well as nutrition. Factors that have influenced consumer's lifestyle as well as their eating habits are based on demographics, socio-economics, culture, politics and environment. These influential factors are reflected in the new trends of food processing, including new developments of techniques and methodologies for the manufacturing, preservation, and monitoring of food. Consumer's desires are oriented towards new foods that are convenient,easy to preserve and promote health and well being. Consumers are also more aware of the need for safer products, increasing the demand of fresh or minimally-processed food devoid of synthetic chemical preservatives.

Another important trend is the consumer demand for healthy foods, *i.e.*, foods that are not only good from a nutritional perspective, but are health promoting (*i.e.*, functional foods or nutraceutical). Other important issues are those related to the health and environmental impact of food processing. Consumers tend to support safer and environmentally benign technologies in the manufacture of new food products. To address the need for safer and healthier foods or compete for consumers acceptance, manufacturers are exploring new food processing and preservation methods. Hence, new trends in food science and nutrition are very much essential so as to be updated with the knowledge regarding existing changes in food science, food processing and food technology. In this chapter, some of the important trends are presented as below :

a) Irradiation

Food irradiation as defined by the International Atomic Energy Agency, is a process involving the exposure of food, either prepackaged or in bulk, to gamma-rays, or electrons in a special room, for a specific duration. The use of gamma-rays to preserve foods by destruction of micro-organisms or inhibition of biochemical changes is called irradiation. Irradiation (cold pasteurization) of food is achieved by exposing the product to a source of ionizing energy. It is a physical means of food processing that involves exposing pre-packaged or bulk foodstuffs to gamma rays (Cobalt-60), X-rays, or electrons.

b) Evapouration

Evaporation refers to the process of heating liquid to the boiling point to remove water as vapour. Because milk is heat sensitive, heat damage can be minimized by evaporation under vacuum to reduce the boiling point. The basic components of this process consist of:

- Heat-exchanger
- Vacuum
- Vapour separator
- Condenser

The heat exchanger is enclosed in a large chamber and transfers heat from the heating medium, usually low pressure steam, to the product usually via indirect contact surfaces. The vacuum keeps the product temperature low and the difference in temperatures high. The vapour separator removes entrained solids from the vapours, channelling solids back to the heat exchanger and the vapours out to the condenser. It is sometimes a part of the actual heat exchanger, especially in older vacuum pans, but more likely a separate unit in newer installations. The condenser condenses the vapours from inside the heat exchanger and may act as the vacuum source (Fig. 14).

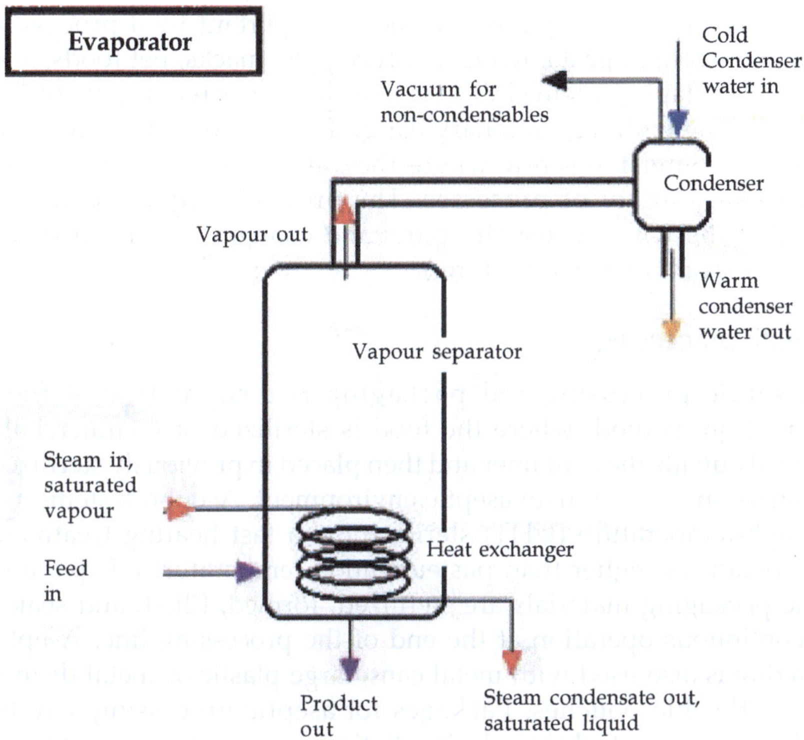

Fig. 14: Method of evaporation

c) Extrusion

Food extrusion is a form of extrusion used in food processing. It is a process by which a set of mixed ingredients are forced through an opening in a perforated plate or die with a design specific to the food, and is then cut to a specified size by blades. The machine which forces the mix through the die is an extruder, and the mix is known as the extrudate. The extruder consists of a large, rotating screw tightly fitting within a stationary barrel, at the end of which is the die.

Extrusion enables mass production of food via a continuous, efficient system that ensures uniformity of the final product. Food products manufactured using extrusion usually have a high starch content. These include some pasta, breads (croutons, bread sticks, and flat breads), many breakfast cereals and ready-to-eat snacks, confectionery, pre-made cookie dough, some baby foods, full-fat soya, textured vegetable protein, some beverages, and dry and semi-moist pet foods.

Extrusion processing has become an important food process in the manufacture of pasta, ready-to-eat cereals, snacks, pet foods, and textured vegetable protein (TVP). An extruder consists of tightly fitting screw rotating within a stationary barrel. Preground and conditioned ingredients enter the screw where they are conveyed, mixed, and heated by a variety of processes. The product exits the extruder through a die where it usually puffs and changes texture from the release of steam and normal forces.

d) Aseptic processing

Aseptic processing and packaging is a sophisticated food preservation method, where the food is sterilized or commercially sterilized outside the container and then placed in previously sterilized containers and sealed in an aseptic environment. Aseptic systems use ultrahigh-temperature (UHT) sterilization, a fast heating treatment at temperatures higher than pasteurization temperatures. Paper and plastic packaging materials are sterilized, formed, filled, and sealed in a continuous operation at the end of the processing line. Aseptic packaging is also used with metal cans, large plastic or metal drums, or large flexible pouches. Packages for aseptic processing may be sterilized using heat, chemicals, irradiation, or a combination of these methods.

Aseptic processing is the process by which a sterile (aseptic) product (typically food or pharmaceutical) is packaged in a sterile container in a way that maintains sterility.

Sterility is achieved with a flash-heating process (temperature between 195 and 295 °F (91 to 146 °C)), which retains more nutrients and uses less energy than conventional sterilization techniques such as retort or hot-fill canning. Pharmaceutical Sterile processing includes use of clean rooms, bacteria retaining filters, dry or steam heat. Aseptic food preservation methods allow processed food to keep for long periods of time without preservatives, as long as they are not opened. The aseptic packages are typically a mix of paper (70%), polyethylene (LDPE) (24%), and aluminium (6%), with a tight polyethylene inside layer. Sterile pharmaceuticals are usually packaged in plastic or glass. Together these materials form a tight seal against microbiological organisms, contaminants, and degradation, eliminating the need for refrigeration.

e) Functional foods

These foods are so designed that the additional health claims are added to the food itself and need not be taken as a separate supplement. These foods have been either enriched, with a deficient nutrient or fortified with extra processing have been restored. Additional components or ingredients are added to ordinary food to give a physiological benefit. Foods which are consumed daily should be enriched with tee /1, which re-date a biological process which can help prevent or control chemical components disease.

f) Antioxidants

Antioxidants are compounds that prevent oxidation. Oxidation is the process that turns cut apples brown. Oxidation leads to degenerative changes in our body, *i.e.*, it contributes to the breakdown of body cells as we age. Antioxidants, such as vitamin C, vitamin E, beta-carotene, and selenium, help protect against free radical damage. They scavenge free radicals and protect body cells against cancer. They prevent atherosclerosis and coronary artery diseases (CAD). The sources of antioxidants are listed below:

1. Vitamin E : Soya oil, sunflower oil, almonds, spinach, and mint
2. Vitamin C : Amla, guava, green leafy vegetables, all citrus fruit, papaya, tomato, cabbage, and capsicum
3. Beta-carotene: All green leafy vegetables, and all yellow, orange, and red fruits and vegetables
4. Selenium : Whole grains, whole pulses, green leafy vegetables, and cauliflower
5. Non-nutrient antioxidants : Phenolic compounds, flavonoids, and isoflavones present in beans, cloves, oats, tea, coffee, grapes, turmeric, mustard, red wine, etc.

   We need to consume much more than the RDA for antioxidant effect.

g) Trans fatty acids

Fatty acids which contain a double bond can exist in either of two geometric isomeric forms: cis form or trans form. Unsaturated fatty acids in foods are usually in the cisconfiguration,*i.e.*, the hydrogen atoms at the double bond are on the same side of the molecule. In the trans configuration, the hydrogen atoms at the double bond are on the opposite side. The trans molecule assumes a saturated fatty acid like configuration. The change in configuration alters the physical property of the oils, changing liquid oils into semi-solid to solid fats. Oleic acid, the cis isomer, is a liquid oil while elaidic acid, its trans isomer is solid. Trans-fatty acids are commercially introduced agents created by partial hydrogenation of the essential fatty acid, thereby reducing their content in the fat.

Their use has been eliminated from retail fats and spreads in many parts of the world, but are still used in baked goods and deep-fat fried fast foods. They are popular in the food industry because they have a high melting point compared to the cis-form and help in making crisper puff pastry and other baked products. However, clinical studies have demonstrated that they are more atherogenic than saturated fatty acids. They increase the LDL cholesterol or bad cholesterol and decrease the HDL cholesterol or good cholesterol affecting the ratio of LDL to HDL cholesterol thereby increasing the risk of coronary heart diseases. Apart from industrially hardened oils, dairy and meat fats are also a source of trans fatty acids.

The Food and Agriculture Organization (FAO), World Health Organisation (WHO) joint committee has recommended that less than 1 per cent energy should come from trans fatty acids, *i.e.*, on a 2,000 kcal diet, trans fatty acid intake should be less than 2g per day. The United States Food and Drug Administration (USFDA) has made it mandatory to mention trans fatty acids content on all processed food. The use of trans fatty acids is banned in some countries.

Sources

Major source of trans fatty acids are partially hydrogenated vegetable oils. They are preferred because they have a long shelf-life, are stable at high temperature of baking and frying and the degree of hydrogenation can be customized for different uses in bakeries. Trans fats are widely used in bakery products, such as biscuits, puff pastry, kharis, cakes, cream fillings and crunchy cookies, deep-fried fast foods, packaged savoury snacks.

The type of fats consumed in the diet is more important than the total amount of fat consumed. While the intake of saturated fatty acids need to be reduced in the diet, both mono- and polyunsaturated fatty acids need to be increased and the total fat intake should be restricted to 30 per cent of total energy for greater cardiovascular protection. This can be achieved through appropriate admixtures of different oils or through genetic modification of oilseed crops.

h) Nutraceuticals

It is now a well-known fact that apart from proteins, carbohydrates, fats, vitamins, minerals, and water present in our food, there are a number of important non-nutritive chemical components which play a vital role in maintaining optimal health. Traditional forms of medicine have been used all over the world since time immemorial for prevention and treatment of disease and for vitality, and longevity. The healing effects of food have been advocated by Hippocrates, the father of Western medicine. Evidence from ancient civilizations have suggested that people were aware of the medicinal benefits of food for healing and curing ailments. Ayurveda has mentioned the benefits of food for therapeutic purpose.

The modern nutraceutical market has its beginnings several thousand years ago. Since the 1980s, it has gained momentum because of new research stressing on the preventive and curative properties of food. Today, nutraceuticals form part of the daily diet of peoplein developed and developing countries worldwide. Let us understand the meaning of this term.

There are many components in food apart from the six nutrients which play a vital role in the link between food and health. An increase in life span and change in lifestyle with the advent of technology have been the two main reasons for interest in non-nutritive components. These chemical components derived from plant sources, our food and micro-organism have shown to possess medicinal benefits and can promote long term health and well-being. These chemical components which are not classified as nutrients are called nutraceuticals. The term 'nutraceuticals has been derived from the words 'nutrition' and 'pharmaceutical' and is now used for food or food products that provide health and medical benefits, *i.e.*, they are used to prevent and treat disease. Following are the types of nutraceuticals:

1. Prebiotics and probiotics
2. Antioxidants
3. Phytochemicals

Definition

Nutraceuticals is a large umbrella term used to describe any product derived from food sources that provides extra health benefits in addition to the basic nutritive value of the food. They claim to improve health, prevent chronic disease, delay the aging process and increase life expectancy. Nutraceuticals are available in many different forms. For example, they may be isolated or purified from foods and are generally sold in medicinal forms. They may also be added to foods we consume everyday.

Table 27: List of nutraceuticals

Health claim	Chemical component	Source
Antioxidants	Flavonoids, resvervatol Naringin Carotenoids Polyphenols Alyll sulfides Anthocyanins	Red grapes Citrus fruits Carrots, Tea wine, dark chocolate onions and garlic Berries
Cholesterol lowering	Solube dietary fiber	Psyllium seed husk
Cancer prevenytation	Sulphoraphane Indoles Capasicin Iso flavonoids Saponins	Broccoli Cabbage peppers Soya beans Beans
Anti-inflammatory	Alpha linolenic acid	Flaxseeds, olive oil
Antibacterial effect	Allicin Paroanthocyanidins	Garlic oil Cranberries

Nutraceuticals are broadly classified as :

1. Dietary supplements
2. Functional foods
3. Phytochemicals

Some examples of nutraceuticals are given below:

i) Prebiotics and probiotics

While probiotics are the beneficial bacteria found in the intestine, prebiotics are special indigestible carbohydrates known as oligosaccharides that feed probiotic bacteria and support their growth and colonization in our gut.

a) Prebiotics : Prebiotics are non-digestible food ingredients found in plants, which stimulate the growth and/or activity of bacteria in our digestive system. They mainly include carbohydrates, such as oligosaccharides and soluble dietary fiber, which stay intact during cooking and baking. Because of this property, they can safely be used in cooked and baked food items.

They act as a functional food are beneficial to health. The non-digestible carbohydrates allow microorganisms to grow, as they reach the intestine unaffected by the digestion process. Prebiotic fiber is selectively fermented in the gut and brings about changes in the composition and activity of microorganisms in our gastrointestinal tract.

The main prebiotics are the Fructooligosaccharides (FOS), Inulin, and Oligofructose and galactooligo saccharides (GOS) which are added to processed foods some are fermented slowly, *e.g.*, the longer chain prebiotic inulin, while others are ended more rapidly, e.g., oligofructose.

Prebiotics increase the number and activity of bifidobacteria and lactic acid bacteria in the gut and increase resistance to invading pathogens. These groups of bacteria have several of beneficial effects on humans, such as improving digestion, enhancing absorption of mineral and strengthening of immune system.

Sources of prebiotics

No plant or food is a prebiotic by itself. Foods contain prebiotics in varying quantities. Sources include whole grains, specially, oats, wheat bran, barley, psyllium, soya beans, and products, pulses, flax seeds, sunflower seeds, fenugreek, carrots, citrus fruits and chicory root are richest sources of inulin, while wheat, almonds and honey contain smaller quantities. Other plant-based foods, such as bananas, onions, leeks, garlic, and tomatoes, contain lesser amount of inulin. Prebiotic fiber supplements are added to processed foods to increase their health benefits. Everyday foods, such as cereals, biscuits, bread, health drinks, yoghurt, and table spreads, contain prebiotics.

Daily allowances: To maintain good digestion, an intake of 6 g of prebiotic has been suggested for healthy individuals and at least 15 g per day to individuals suffering from active digestive disorders.

Beneficial effects of prebiotic fiber

Prebiotics promote the growth of probiotics or beneficial bacteria. Bacteria such as *Clostridium and Bacteroides* species produce short chain fatty acids (SCFA). Prebiotics lead to the increased production of SCFA which creates a medium conducive to good digestive health because SCFAs nourish the colon walls, create a slightly acidic medium which helps in reducing the number of sulphate producing bacteria which form H_2S gas. When the diet is low in fiber over a long period of time, SCFA production reduces. Beneficial effects are as follows:

1. An increase in calcium and other mineral absorption
2. Increased immune system effectiveness
3. Regulation of bowel pH
4. Decrease in risk of colorectal cancers
5. Reduced risk of Crohn's disease and ulcerative colitis
6. Reduction in hypertension.

b) Probiotics: Probiotic bacteria are also known as beneficial bacteria or friendly bacteria and are nature present in some of our foods or are added to processed foods for their health benefits. Probiotics are the live micro-organisms present in the food we eat. They are not affected by our digestive process, hence live, healthy bacteria directly reach our large intestine or colon where they grow and form colonies. These healthy bacteria in the intestine can combat harmful bacteria and provide a number of health benefits which have been discussed under prebiotics.

Prebiotics are beneficial because they stimulate the growth of healthy bacteria such as species of *Bifidobacteria and Lactobacilli* in the gut. They are many types of probiotic bacteria but *Lactobacilli and Bifidobactirea* are present in the largest numbers and are also popular probiotics added as a supplement to various foods.

The following bacteriae most commonly found in supplements and capsules :

- *Lactibacillus acidophilu*
- *Lactibacillus rhamnosus*
- *Lactibacillus bulgaricus*
- *Lactibacillus plantarum*
- *Lactibacillus Casei*
- *Lactibacillus sporogenes*
- *Clostridium butyricum*
- *Bifidobacteria bifidum*
- *Bifidobacteria longum*
- *Bifidobacteria infantis*
- *Streptococcus thermophiles*
- *Streptococcus faecalis*
- *Bacillus mesentericus*

Antibiotics are administered to destroy pathogenic organisms. Along with pathogens, they also destroy the normal intestinal flora. Probiotic bacteria are administered in the form of capsules along with anti-fungal or antibacterial therapy to restore the proper balance of organisms in the gut. They prevent colonization by other organisms in the gut. Recent research indicates that the gut flora could be related to development of food and airborne allergies and improving the gut flora could reduce the severity and number of allergies as well.

Sources of probiotics

Sources of probiotics include fermented milk products, such as curds, yoghurt, and butter-milk and their variations. *Lactobacillus casei* is added to Yakult, a probiotic yoghurt drink. Probiotics are added along with prebiotics to everyday products as well as special health drinks, powders, and capsules. They work best in combination.

Bifidobacteria:They prevent and treat antibiotic induced diarrhoea. Bifidobacteria synthesize B-complex vitamins. They help regulate normal bowel movements and prevent colonization of pathogenic bacteria by producing acids which maintain the pH balance.

Lactobacilli: They help in digestion of lactose and dairy products. They acidify the intestinal tract and improve nutrient absorption. The acidic pH prevents pathogens. Like *Bifidobacteria*, they prevent and treat antibiotic induced diarrhoea.

Lactobacilli casei* and *Lactobacilli salivarius: They significantly inhibit the growth of peptic ulcer causing bacteria *Helicobacter pylori (H. pylori).*

j) Role of phytochemicals

A number of newly launched products have included phytochemicals ('phyto' is a Greek word meaning 'plant') in their composition. Phytochemicals are biologically active, naturally-occurring chemical components in plant foods. They are beneficial to health as they reduce the risk of cancer and other chronic diseases. They influence hormonal and enzymatic activity but their exact mechanism is not yet known. These chemical compounds are not proteins, carbohydrates, fats, vitamins, minerals or fiber but are non-nutrient chemicals found in plants, such as the plant pigments, flavonoids, carotenoids, anthocyanin, lycopene, indoles, terpenes, sulfides, and phytoestrogens. These phytochemicals were earlier thought to be 'associated food factors' might be as nutritionally important as vitamins and minerals. Today these 'associated food factors' are called phyto nutrients. They may provide many added nutritional benefits and can help reduce our risk of developing chronic health conditions, such as heart diseases, stroke, cataract, Parkinson's disease, or even cancer as indicated in the Fig. 15 below:

They are present in abundance in parsley, broccoli, alfalfa, spinach, carrots, Acerola cherries, kelp, soya beans, turmeric, yellow corn, tomatoes, cranberries, blueberries, garlic, onions, flaxseeds, walnuts, *etc.* Since these phytochemicals are found widely distributed in fruits and vegetables, a generous intake of fruits and vegetables containing phytochemicals is a better option instead of consuming phytochemicals in the form of nutrient supplements or added in multivitamin pills.

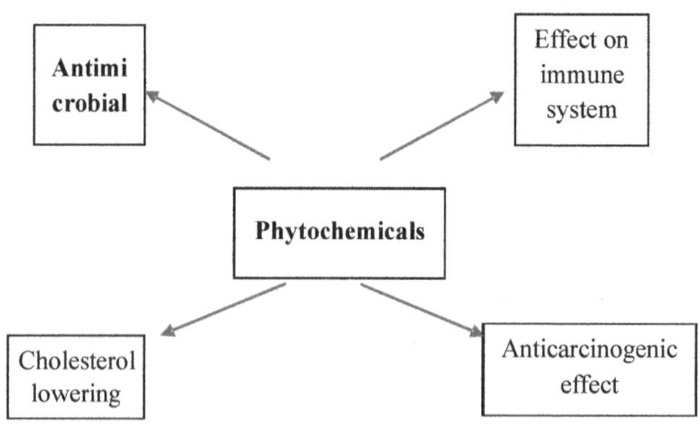

Fig. 15: Beneficial effects of Phytochemicals

k) Nutrition labeling

Nutrition labeling is intended to provide information on regulations related to labelling. Nutrition labels describe the nutrient content of a food and are intended to guide the consumer in food selection. The nutrition information provided must be selected on the basis of consistency with dietary recommendations.

Selection of the specific nutrients or food components to be listed should further take into account label space, the analytical feasibility of measuring the particular nutritional component within the food matrix, and the relative costs of such analyses. Nutrition information provided on labels should be truthful and not mislead consumers. At the same time, labeling regulations should provide incentives to manufacturers to develop products that promote public health and assist consumers in following dietary recommendations. It is likely that in many countries, there would be some segments of the population that would benefit from information about the composition of foods. In these cases, countries should consider the need to provide for appropriate labeling and its presentation relative to existing guidelines and approaches.

As nutrition-labelling efforts have evolved, different approaches and legal requirements have been established. These create difficulties in developing and harmonizing nutrition information listings, which have broad international applications. For these reasons, the Codex guidelines on nutrition labelling play an important role to provide guidelines These guidelines are based on the principle that no food should be described or presented in a manner that is false, misleading or deceptive. The guidelines include provisions for voluntary nutrient declaration, calculation and presentation of nutrient information.

Nutrition labeling by itself cannot solve nutrition problems. It should be seen as one of the elements of nutrition policy and should be envisaged in the larger perspective of consumer education, which in its turn is part of an overall development policy.

Principles for nutrition labelling

a) Nutrient declaration

Information supplied should be for the purpose of providing consumers with a suitable profile of nutrients contained in the food and considered to be of nutritional importance. The information should not lead consumers to believe that there is exact quantitative knowledge of what individuals should eat in order to maintain health, but rather to convey an understanding of the quantity of nutrients contained in the product. A more exact quantitative delineation for individuals is not valid because there is no meaningful way in which knowledge about individual requirements can be used in labelling.

b) Supplementary nutrition information

The content of supplementary nutrition information will vary from one country to another and within any country from one target population group to another according to the educational policy of the country and the needs of the target groups.

c) Nutrition labeling

Nutrition labels describe the nutrient content of a food and are intended to guide the consumer in food selection. The nutrition information provided must be selected on the basis of consistency with dietary recommendations.

Codex guidelines on Nutrition labelling

To ensure that nutrition labelling is effective

- In providing the consumer with information about a food so that a wise choice of food can be made;
- In providing a means for conveying information of the nutrient content of a food on the label;
- In encouraging the use of sound nutrition principles in the formulation of foods which would benefit public health;
- In providing the opportunity to include supplementary nutrition information on the label.

To ensure that nutrition labeling does not describe a product or present information about it which is in any way false, misleading, deceptive or insignificant in any manner.

To ensure that no nutritional claims are made without nutrition labelling.

❑❑❑

References

Anonymous, 2010. Nutrient requirements and recommended dietary allowances for Indians, National Institute of Nutrition, ICMR, Hyderabad.

Dietary guidelines for Indian, 1998, ICMR, NIN, Hyderabad.

Gajalakshmi, R. 2015, Nutrition Science, CBS Publishers and distributors Pvt. Ltd. FSSAI, 2010, Training manual for Food Safety Regulations, Vol.I-Introduction to food and food processing.

Gopalan C., Ramasastri, B.V. and Balasubramanian, S.C., 2010, Nutritive value of Indian foods, National Institute of Nutrition, ICMR, Hyderabad.

Harbans Lal, 2012, Food and Nutrition 2nd Edition, CBS Publishers and distributors Pvt. Ltd, New Delhi.

Javier Se~Norans, f. Javier Se~Norans, Elena IB'A~Nez, and Alejandro Cifuentes, New Trends in Food Processing, *Critical Reviews in Food Science and Nutrition,* 43(5):507-526.

Mudambi, S.R. and Rajagopal, M.V. 2007, Fundamentals of foods Nutrition and diet therapy, 5th edition, New age publications.

Neena Joshi, Revanna, M.L. and Usha Ravindra, 2015, Nutrition Management and Food safety, PGDA205 Self learning study material, Published by Distance Education Unit, Directorate of Extension, UAS, Bangalore.

Penchalaraju, M., Kavitha, C. and Kirthy Reddy, M. 2015, Food Processing Novel Technologies, Kalyani Publishers, Bengaluru, First Edition. Principles of Food Science,MFN-008, Study material, Indira Gandhi National Open University School of Continuing Education.

Puja Dudeja, Amarjeet Singh and Sukhpal Kaur, 2016, Food Safety Implementation, First Edition.

Shakuntala Manay, N., Shadakshara Swamy, M., 2005, Foods Facts and Principles., Second Edition. New Age international publishers, New Delhi.

Sharma D.C. and Devanshi Sharma, 2015, First Edition, Nutritional Biochemistry, CBS Publishers and distributors.

Srilakshmi, B. 2000. Nutrition Science, New age Publishers.

Sunetra Roday, 2016, Food Science and Nutrition, Oxford University Press, New Delhi, Second edition.

Vijaya Khader, 2013, Food Science and Technology, Indian Council of Agricultural Research, New Delhi.

Websites

http://www.encyclopedias3.com/encyclopedia-of- food-and-culture-volume-1/3197-physiological-fuel- values-of-foods.html

https://www.google.co.in/imgres?imgurl=https://ka-perseus mages.s3.amazonaws.com/71225d815cafcc09102504abd

f4e10927283be98.png&imgrefurl=https://www.khanacademy.org/science/biology/macromolecules/proteins-and-amino-acids/a/orders-of-protein-structure

https://www.google.co.in/imgres?imgurl=http://chemistry.elmhurst.edu/vchembook/images/568denaturation.gif&imgrefurl=http://chemistry.elmhurst.edu/vchembook/568denaturation.

https://en.wikipedia.org/wiki/Food_extrusion

https://www.ncbi.nlm.nih.gov/pubmed/378548

https://www.sciencedirect.com/topics/agricultural-and-biological-sciences/aseptic-processing

https://en.wikipedia.org/wiki/Aseptic_processing

https://www.uoguelph.ca/foodscience/book-page/evaporation

https://timesofindia.indiatimes.com/india/India-ranks-below-China-Pak-in-global-hunger-index/articleshow/6728259.cms

Model Questions

I. Define the following:

1. Malnutrition
2. BMI
3. RDA
4. Nutritional Status
5. Health
6. Metabiosis
7. Relative density
8. Surface tension
9. Bioactive compounds
10. Evaporation
11. Non-essential amino acid acids.

II. Choose the Correct answer

1. The temperature of freezing is __________

   a) -18°C b) -19°C

   c) -20°C d) -21°C

2. Lactobacillus species is used in __________products.

   a) Fermented dairy b) Vegetable fermentation

   c) Meat fermentation d) All the above

3. Yogurt is industrially produced fermented milk using a particular __________ as a Starter

   a) Yeast b) Mold

   c) Bacterium d) Fungi

4. The natural toxin present in Soy beans

   a) Pepsin b) Trypsin

   c) Lypase d) Amylase

5. Nutrients in food are better retained by controlled processing conditions:

   a) Autoclaving b) Freeze drying

   c) Controlled heat d) All the above

6. Processed margarine is a substitute for:

   a) Cheese b) Butter

   c) Vanaspathi d) Fresh cream

7. Processed foods are often enriched with:

   a) Protein b) Fats

   c) Vitamins & Minerals d) CHO

8. The freezing method in which liquid N_2 is used and called as:

   a) Plate freezer b) Blast freezer

   c) Immersion freezer d) Cryogenic freezing

9. The foods under cook-freeze has a shelf-life of:

   a) 1-2 months b) 2-3 months

   c) 3-6 months c) 6-12 months

10. Holding pasteurization method is commonly used for:

    a) Fermented beverages b) Milk products

    c) Fruit juices d) Ice- creams

11. The thermal death of microorganism in dry air is due to:

    a) Oxidation b) Condensation

    c) Asepsis d) Sterilization

12. Benefits of fermentation are:

    a) Flavor changes b) Texture changes

    c) Nutritional changes d) All the above

III. Match the following

	A	B	Ans
1.	Ground nut	a) 4% Acetic acid	**3**
2.	Wine	b) Olive Oil	**11**
3.	Vinegar	c) Retardation of microorganism	**6**
4.	Lactobacilli	d) Air line and Institution catering	**7**
5.	Plant equipment	e) Dry fruits	**8**
6.	Asepsis	f) Acidity	**4**
7.	Cook - chill system	g) < 130°C	**9**
8.	SO_2	h) Oxidation	**5**
9.	HTST	i) Aflatoxin	**1**
10.	Sodium Benzoate	j) Must	**2**
11.	Unsaturated fatty acid	k) Tomato sauce	**10**
12.	UHT	l) > 150°C	**12**
13.	Probiotics	m) Swiss cheese	**22**
14.	Tempeh	n) Ketabolism	**17**
15.	Antioxidant	o) Stored in the muscle & liver	**19**
16.	Intracellular fluids	p) Precursor of vitamin D	**20**
17.	Breakdown of tissue protein	q) Lactobacillus	**12**
18.	Hemicellulose	r) Folic acid	**21**
19.	Glycogen	s) Potassium	**16**
20.	7Dehydrocholestrol	t) Polysaccharides	**18**
21.	Megaloblastic anemia	u) Fermented soybean product	**13**
22.	Propionibacterium	v) Tocopherol	**15**

IV. Answer the following questions:

1. List the factors affecting growth and survival of micro-organisms.
2. Write the objectives of food processing.
3. Write the classification of food preservatives.
4. Write the classification of CHO with examples.
5. Write ten functions of dietary fibre.
6. What are the methods/tests to estimate the quality of protein?
7. List any four bioactive components.
8. Write the properties/ characteristics of lipids.
9. Write a note on water activity.
10. Explain about the role of phytochemicals.
11. Discuss about the food spoilage micro-organisms.
12. What is fermentation and its application in food preservation? Explain briefly.
13. Discuss briefly about the functions of folic acid.
14. Write a short note on irradiation.
15. Discuss about water soluble vitamins.

V. Answer the following questions Briefly

1. Discuss on etiology, symptoms and dietary managements of kwashiorkor and marasmus in detail.
2. Explain about the principles and methods available for food processing.
3. Explain about the food pyramid menu planning and factors influencing menu planning.
4. Explain about the symptoms of vitamin C deficiency, sources and RDA of Vitamin C.
5. Discuss about the importance and significance of nutritional labelling.

❑❑❑